NONPARAMETRIC STATISTICS FOR HEALTH CARE RESEARCH

NONPARAMETRIC STATISTICS FOR HEALTH CARE RESEARCH

Statistics for Small Samples and Unusual Distributions

Marjorie A. Pett

SAGE Publications
International Educational and Professional Publisher
Thousand Oaks London New Delhi

For information:

 SAGE Publications, Inc.
2455 Teller Road
Thousand Oaks, California 91320
E-mail: order@sagepub.com

SAGE Publications Ltd.
6 Bonhill Street
London EC2A 4PU
United Kingdom

SAGE Publications India Pvt. Ltd.
M-32 Market
Greater Kailash I
New Delhi 110 048 India

Printed in the United States of America

Library of Congress Cataloging-in-Publication Data

Pett, Marjorie A.
  Nonparametric statistics for health care research: statistics for
small samples and unusual distributions / author, Marjorie A. Pett.
     p. cm.
  Includes bibliographical references and index.
  ISBN 0-8039-7038-2 (cloth). — ISBN 0-8039-7039-0 (pbk.)
  1. Medicine—Statistical methods. 2. Nonparametric statistics.
  I. Title.
  R853.S7P48 1997
  610'.72—dc20   96-35711

01 02 03 10 9 8 7 6 5 4

Acquiring Editor:	Dan Ruth
Editorial Assistant:	Jessica Crawford
Production Editor:	Diana E. Axelsen
Production Assistant:	Denise Santoyo
Typesetter/Designer:	Christina M. Hill/Andrea D. Swanson
Indexer:	Virgil Diodato
Cover Designer:	Ravi Balasuriya
Cover Illustration:	Mark Pett
Print Buyer:	Anna Chin

Contents

Preface	7
Organization of the Text	8
Those Who May Find the Text Useful	9
Acknowledgments	**11**
1. Overview of Nonparametric Statistics	**13**
Common Characteristics of Parametric Tests	13
Development of Nonparametric Tests	15
Characteristics of Nonparametric Statistics	16
Use of Nonparametric Tests in Health Care Research	17
Some Common Misperceptions About Nonparametric Tests	18
Types of Nonparametric Tests	19
2. The Process of Statistical Hypothesis Testing	**20**
Choosing Between a Parametric and a Nonparametric Test	25
3. Evaluating the Characteristics of Data	**30**
Characteristics of Levels of Measurement	30
Assessing the Normality of a Distribution	35
Dealing With Outliers	49
Data Transformation Considerations	52
Examining Homogeneity of Variance	54
Evaluating Sample Sizes	54
Reporting Testing Assumptions and Violations in a Research Report	57
Summary	57
4. "Goodness-of-Fit" Tests	**59**
The Binomial Test	59
The Chi-Square Goodness-of-Fit Test	69
The Kolmogorov-Smirnov One-Sample Test	79
The Kolmogorov-Smirnov Two-Sample Test	87
Discussion	93
5. Tests for Two Related Samples: Pretest-Posttest Measures for a Single Sample	**95**
The McNemar Test	96

The Sign Test 105
The Wilcoxon Signed Ranks Test 112
Discussion 121

6. Repeated Measures for More Than Two Time Periods
   or Matched Conditions 122
   Cochran's Q Test 123
   The Friedman Test 131
   Summary 145

7. Tests for Two Independent Samples 146
   The Fisher Exact Test 147
   The Chi-Square Test for Two Independent Samples 157
   The Wilcoxon-Mann-Whitney U test 169
   Summary 179

8. Assessing Differences Among Several Independent Groups 181
   The Chi-Square Test for k Independent Samples 182
   The Mantel-Haenszel Chi-Square Test for Trends 198
   The Median Test 204
   The Kruskal-Wallis One-Way ANOVA by Ranks 212
   Summary 223

9. Tests of Association Between Variables 225
   The Phi Coefficient 225
   Cramér's V Coefficient 232
   The Kappa Coefficient 237
   The Point Biserial Correlation 248
   The Spearman Rank-Order Correlation Coefficient 255
   Kendall's Tau Coefficient 265
   Summary 274

10. Nonparametric Statistics: The Current State of the Art 276
    Currently Available Nonparametric Statistical Procedures 276
    The Limitations of Available Nonparametric Statistics 278
    Finally—The $64,000 Question 280

Appendix: QBASIC Routine for Subpartitioning an
   $r \times k$ Contingency Table 281

References 283

Index 293

About the Author 307

Preface

Consider this scenario: You have a wonderful idea for a health care intervention that you believe will have a strong positive impact on a particular client outcome. You work in a setting in which such an intervention can be undertaken feasibly. There is even an available comparison group that could receive traditional patient care, *and* there is a pre- and posttreatment measure that is a reliable and valid assessment of the outcome in which you are interested. What more could a researcher in health care ask for? Your sample size? Well, not as large as you would like—there are only 10 patients per group—but, realistically, given the setting, time constraints, and lack of funding, this is about as large as this sample is likely to be.

You undertake this study, collect the data, choose your statistical package, and are ready to determine the most appropriate statistics to use in this study. What statistics should you use? Not a problem, you think. Both your textbook on statistics and your software package indicate that there are plenty of tests to choose from. There are ANOVAs, ANCOVAs, and MANOVAs. There are also *t* tests, Pearson correlations, and multiple regression. You recall that all these tests are labeled *parametric* and that they are based on some important assumptions, not the least of which is an adequate total sample size, a certain level of measurement, and a normal distribution of the dependent variable. Your study has violated a number of these assumptions.

Does this sound familiar? It should. This problem is relatively common in the health care research literature. That is, the research was undertaken on a limited budget, using a small sample of convenience, in a health care setting by a researcher whose primary interests are improving patient care and facilitating more satisfactory patient outcomes. Given these necessary limitations, it is inevitable that much of health care research is beset with potentially serious violations of the assumptions of parametric statistics. Unfortunately, however, most statistics textbooks at both the undergraduate and graduate levels tend to focus on parametric statistics, reserving only a few pages for their nonparametric counterparts. Most researchers in health care, therefore, have had limited exposure to alternatives to parametric tests. It is the selection and interpretation of the most appropriate statistical tests, however, that enable researchers to assess the effectiveness of a particular intervention. It is critical, therefore, that researchers be knowledgeable about the use and potential for misuse of parametric tests in health care research, know when to use such statistical techniques, and be aware of the availability of *nonparametric* alternatives when assumptions of parametric tests have been violated.

The purpose of this book is to present practical information concerning some of the most commonly used nonparametric statistical techniques that are available in the most frequently used statistical computer packages. The book's intention is to help you to understand when a particular nonparametric statistic would be used appropriately, to learn how to generate and interpret the computer printouts resulting from the application of this statistic, and to present the results of the analysis in table and text format. For the sake of consistency and because of its popularity, SPSS for Windows (Norusis, 1995a) is the statistical package that will be used throughout this text. I realize that users of other equally adequate statistical packages (e.g., BMDP, MINITAB, and SYSTAT) may initially view this conscious choice of a single environment to be disappointing and potentially restricting. As you read the text, I think that you will find the information provided will better enable you to select the most appropriate nonparametric test for your data. However, despite our differences in choice of statistical packages, the approach to examination and interpretation of the generated computer printouts will not change.

Organization of the Text

This book consists of 10 chapters. Chapter 1 presents an overview of the development of nonparametric tests, a comparison of the characteristics of parametric and nonparametric tests, the use and common misperceptions of nonparametric statistics in health care research, and their current availability in statistical packages for personal computers. Chapter 2 examines the issues that are crucial in choosing the best statistical test given a particular data set. Included in the discussion are a brief review of the process of statistical hypothesis testing and an outline of criteria useful for choosing the most appropriate statistical test. Chapter 3 presents ways to evaluate the characteristics of data, such as level of measurement, normality of distributions, assessment of outliers, homogeneity of variance, and adequacy of sample size.

Chapters 4 through 9 each examine a particular set of nonparametric statistics using a common presentation format: the purpose of the particular statistic, a research question that could be answered using the statistic, an example of null and alternative hypotheses that would follow from the research question, an overview of the test procedure, a discussion of the test's underlying assumptions and limitations, the SPSS for Windows commands used to generate the statistic, interpretations of the resulting computer printout, suggested ways to present the results in tabular and written format, and examples from the published research literature from a variety of health care disciplines (e.g., exercise and sports science, health education, medicine, nursing, psychology, and social work) that have used the statistic in their analyses. Substantive references that provide further information about the statistic are located at the end of the book.

Chapter 10 summarizes the current availability of these nonparametric alternatives to parametric tests and presents a summary reference guide to the nonparametric statistics presented in the text along with an identification of their parametric alternatives. Although these chapters may appear to be a "cookbook" approach to the applications of nonparametric statistics, it is intended that the reader will become familiar with the use, logic, assumptions, and interpretations of these applications.

It should be noted that a critical assumption of *all* parametric and nonparametric tests is that the data be obtained from a random sample. Because of its universality, the assumption of random sampling will not be repeated for all the nonparametric tests discussed in this text. This is, however, an important assumption, especially when assessing generalizability of the obtained results. Without random selection, the ability of the researcher to assess the representativeness of his or her sample may be compromised.

Those Who May Find the Text Useful

This book is intended to be read by researchers, students, and professionals from a variety of settings and disciplines who do not necessarily have strong mathematics backgrounds but who are interested in finding reasonable and practical alternatives to parametric statistics when their data do not meet the assumptions of these tests. For that reason, presentation of mathematical formulas will be kept to a minimum.

This book is written at a level easily understood by people who have had at least a beginning course in statistics and therefore possess some familiarity with the concepts of levels of measurement, statistical hypothesis testing, shapes of distributions, outliers, various parametric statistics, and statistical power. For those readers who are unfamiliar with these topics, a close reading of Chapters 2 and 3 is strongly recommended. Additional resource references are provided in these chapters for further readings on these topics.

Because it is felt that nonparametric statistics have a particular usefulness in health care research, the examples given throughout the text have been directed to that area. Readers who have other research interests, however, should find that much of the information provided also applies to their areas of focus. Finally, it is this author's hope that by the conclusion of this text the informed reader will be better able to answer the critical and somewhat intoxicating question,

Does the Kolmogorov-Smirnov test *really contain vodka?*

Acknowledgments

There are many people to whom I am very much indebted for helping to make this text a reality. First, I would like to thank Deans Linda Amos and Kay Dea and the faculties in the College of Nursing and the Graduate School of Social Work, respectively, at the University of Utah for their patience and support not only for this text but also for our continued relationship. I am appreciative as well to the many graduate students in nursing, social work, exercise and sports science, and health education who were willing to expand their creative thinking and skills to "embrace"— albeit at times reluctantly—the many challenges and potential rewards offered by statistics. Without your support, the elective in nonparametric statistics, and subsequently this text, would not have been a reality. Thanks in particular to Shu Li Chen, Lee Dibble, Lorrie Larsen, Robin Marcus, Sandra Smith, and Sharon Stephens for their helpful critiques and valuable suggestions for improvement of this manuscript.

I am most indebted to my secretary, Darlene Buist, for her cheerfulness and skill in editing this manuscript; to Dan Ruth, the staff, and reviewers at Sage Publications; and to Christine Smedley, in particular, for being in the right place at the right time. Finally, to my immediate family—Art, Mark, and Una—thanks for sharing both the dining room table and other parts of your busy lives with the latest version of this manuscript. Mark, I especially appreciate your having illustrated the cover of this book. To Mom and Dad, unfortunately you died before this text became a reality, but it is to your memory that I dedicate this effort. Godspeed.

SPSS™ is the registered tradement of SSPS Inc., 444 North Michigan Avenue, Chicago, IL 60611-3962. (Phone: 312-329-2400.) SSPS dialogue boxes from SPSS for Windows are printed with the permission of SPSS, Inc.

1 Overview of Nonparametric Statistics

Historically, the most popular statistical inferential techniques that have appeared in the research literature are those that make assumptions about the nature of the populations from which the data are drawn. These techniques are called *parametric* statistics because of their focus on specific parameters of the population, especially the population mean and variance.

Common Characteristics of Parametric Tests

Parametric tests share a number of common characteristics (see Box 1.1). First, it is expected that there is independence of observations except when the data are paired. The data also are expected to be randomly drawn from a normally distributed, or bell-shaped, population of values. This condition is actually a proxy for the *real* assumption that the distributions of the parameters being tested (such as the mean) are normal. It is also expected that the dependent variable being analyzed is measured on at least an interval-level scale. That is, these data are rank ordered and have units or numbers that have equal intervals and whose values share similar meanings (e.g., a person's weight, a score on a depression scale that ranges from 1 to 100, or a preterm infant's heart rate).

Given that these data are assumed to be normally distributed, it is important to make some assessments of the normality assumption. Although not a requirement, to assess normality, a minimum sample size of approximately 30 subjects per group has been recommended. This traditional and somewhat arbitrary sample size recommendation is linked to the Central Limit Theorem, which states that even when the population is

---

Box 1.1 Common Characteristics of Parametric Tests

- Independence of observations, except when the data are paired
- The observations for the dependent variable have been randomly drawn from a normally distributed population
- The dependent variable is measured on at least an interval-level scale; that is, one that is rank ordered and has equidistant numbers, with the numbers sharing similar meaning
- A minimum sample size of approximately 30 subjects per group is recommended
- Data are drawn from populations having equal variances (rule of thumb: one variance cannot be twice as large as the other)
- Usually hypotheses are made about numerical values, especially the mean of a population (μ), for example:

    H_0: $\mu_1 = \mu_2$
    H_a: $\mu_1 \neq \mu_2$

- Other possible requirements: nominal or interval-level independent variable, homoscedasticity, and equal cell sizes

---

nonnormal, the sampling distribution of the mean (a critical parameter for parametric tests) becomes more like the normal distribution as the sample size increases (Hinkle, Wiersma, & Jurs, 1994). Moreover, if comparing two or more groups, these sets of data should be drawn from populations having equal variances or spread of scores. The null and research hypotheses are formulated about numerical values, especially the means of a population. A typical null hypothesis is that the populations of interest share a common mean with regard to the dependent variable of interest (e.g., H_0: $\mu_1 = \mu_2$). The alternative or research hypothesis states that the population means are not the same (e.g., H_a: $\mu_1 \neq \mu_2$).

Some parametric tests have additional requirements. These include assumptions regarding the level of measurement for the independent variables (nominal for ANOVA, at least interval for Pearson product moment correlations), homoscedasticity, and equal cell sizes. *Homoscedasticity* implies that for every level of the independent variable, the dependent variable has a similar variance.

If a parametric test is the statistical technique of choice, the results should be, but unfortunately are not often, presented with this caveat:

If our assumptions concerning the shape of the population distributions are valid, we may conclude that . . .

Because of this common set of assumptions, parametric tests are thought to be more clearly systematized, easier to apply, and supposedly easier to teach, although those of us who have taught or have taken classes in parametric statistics may think otherwise. Parametric tests, therefore, have been extremely popular in the research literature, almost to the point of excluding any other techniques.

Development of Nonparametric Tests

There are alternative tests of statistical inference that do not make numerous or stringent assumptions about the population from which the data have been sampled. These techniques have been called *distribution-free* or *nonparametric* tests. A substantial body of published information exists concerning these tests. Both Savage (1962) and Singer (1979), for example, present bibliographies containing thousands of articles and books about nonparametric statistics.

A common misperception about nonparametric statistics is that they are relative newcomers to the statistics arena. This is not the case. Singer (1979) points out that, historically, parametric and nonparametric statistics appear to have been developed conjointly. In an entertaining overview of the historical development of nonparametric statistics, Singer indicates that the first attempt at hypothesis testing using a statistical test was undertaken by Artbuthnot in 1710. The procedure that was used was the nonparametric sign test. Singer also points out that Artbuthnot was friends with DeMoivre, the man who has been credited with first describing the normal distribution. It appears, therefore, that some of the best friends of statisticians of parametric persuasion are their nonparametric colleagues.

Despite having developed simultaneously, parametric and nonparametric tests have not shared the same popularity. Perhaps the term *nonparametric* implies that there is something lacking in nonparametric statistics, thus contributing to these statistics' second-class status among researchers. Even as early as the 1920s, however, there was concern expressed about the exclusive reliance on the normal distribution in the research literature. Singer (1979) reports that Karl Pearson (1920), the man credited with naming the bell-shaped distribution as *normal*, warned that the normal distribution "has the disadvantage of leading people to believe that all other distributions of frequencies are in one sense or another 'abnormal' . . . that

---

Box 1.2 Common Characteristics of Nonparametric Tests

- Independence of randomly selected observations except when paired
- Few assumptions concerning the population's distribution
- The scale of measurement of the dependent variable may be categorical or ordinal
- The primary focus is on either the rank ordering or the frequencies of data
- Hypotheses are most often posed regarding ranks, medians, or frequencies of data
- Sample size requirements are less stringent than for parametric tests

---

belief is, of course, not justifiable" (Pearson, 1920, p. 25). It is interesting that Pearson should have made that comment more than 75 years ago and yet researchers still seem to hold that belief today.

Characteristics of Nonparametric Statistics

Just because these nonparametric or distribution-free tests do not have the strict assumptions of parametric tests does not mean that nonparametric tests are assumption-free (see Box 1.2). Like parametric tests, nonparametric tests assume independence of randomly selected observations except when data are paired. Unlike parametric tests, however, there are limited assumptions required concerning the shape of the population's distribution. Because of the distribution *free-er* nature of the data (McSweeney & Katz, 1978), the distribution of values for the dependent variable may be very skewed, or nonnormal. That is, distributions can take on any shape and are not limited to the bell shape of the normal distribution. When comparing two or more groups using rank tests, however, the distributions of these values within each group must be similar in shape except for their locations (i.e., medians). The dependent variable may also be categorical or rank ordered (i.e., ordinal). Examples of categorical data are a person's marital status (married, divorced, or single). Ordinal data could be a person's response on a 7-point Likert-type scale to a question concerning his or her current stress level (1 = *not at all stressed* to 7 = *extremely stressed*). Chapter 3 of this text reviews the levels of measurement of variables in greater detail.

In nonparametric tests, the primary focus is either on the *rank ordering* of scores, not their actual values, or on the *frequencies* or classification of data. The hypotheses that are posed, therefore, concern ranks, medians, or frequencies, not population means. Sample size requirements also are less stringent for nonparametric tests. It is not unusual for sample sizes of 20 or less to be reported.

A considerable body of research has indicated that parametric tests are more powerful than nonparametric tests only if the assumptions of the parametric test under consideration have been met. When choosing the most appropriate statistical test, therefore, it is important to examine carefully the extent to which the data to be analyzed adequately meet the test's assumptions. It is also necessary to evaluate the consequences of violating certain of the assumptions underlying the test that the researcher is considering.

No data will perfectly meet all of a test's assumptions. For example, the collected data may have a skewed or flat (i.e., nonnormal) distribution. Although some parametric tests are *robust* and can withstand certain violations of their assumptions, other tests are not so flexible. In the latter situation, the parametric test may be a poor choice in contrast to its nonparametric counterpart. Statistics are tools designed to help the researcher in health care to make informed decisions about the outcomes of potentially important interventions. It is ultimately the client who pays the price for poorly planned and executed statistical analyses.

Use of Nonparametric Tests in Health Care Research

Nonparametric statistics have a high potential for use in health care research. Their acceptance of small sample sizes, use of categorical- or ordinal-level data, and ability to accommodate unusual or irregular sampling distributions make them plausible alternatives to the more stringent parametric tests. Unfortunately, however, although most statisticians might agree on the need for both classes of tests in health care research, the reality is that nonparametric tests continue to be underused in research from a variety of disciplines.

Gaither and Glorfeld (1983), for example, report that in an examination of 1,102 articles that appear in the organizational behavior journals from 1976 to 1981, parametric tests dominated as the tests of choice. Only 169, or 9.3%, of the 1,824 statistical procedures used in these articles were nonparametric. The most common nonparametric procedures applied were the chi-square tests of independence and goodness of fit. These two tests accounted for 92 (54.4%) of the 169 nonparametric tests reported. In an in-depth examination of 100 randomly selected articles, the authors concluded that, at best, there was insufficient evidence reported in the journal articles to indicate that the appropriate choice of parametric tests had been made. They also suggested that, because of the lax reporting, it was difficult to ascertain whether inappropriate choices of parametric tests had been made. Conclusions similar to those of Gaither and Glorfeld were reached by Jenkins, Fuqua, and Froehle (1984) in counseling psychology and Pett and Sehy (1996) in nursing.

There does not appear to be any specific discipline that is free from this trend. Buckalew (1983) suggests that nonparametric techniques are often portrayed as inferior to the more popular parametric tests, both in teaching and in practice. He also argues that these tests deserve greater recognition and use in psychology. Similar arguments have been presented by Lezac and Gray (1984) for neuropsychology, Royeen and Seaver (1986) for occupational therapy, and Harwell (1988, 1990) for psychology and education.

Some Common Misperceptions About Nonparametric Tests

So why is it that, despite their great promise and potential, nonparametric statistics continue to be underused in health care research? Singer (1979) suggests that there are some common misperceptions related to nonparametric statistics on the part of readers, reviewers, and authors of research that may contribute to underuse of these statistics. These misperceptions include fears that readers of manuscripts might not understand the statistics and that the manuscripts might not be accepted by editors or reviewers. Some researchers also perceive that nonparametric statistics are inferior to parametric tests, that such statistics are available only for the simplest of research designs, and that few statistical computer packages contain these statistics.

Such misperceptions appear to be related, in part, to a limited exposure to and training in the wide variety of uses to which nonparametric statistics can be applied. Unfortunately, nonparametric statistics often are relegated to the final pages of a chapter or textbook on statistics. Pett and Sehy (1996) also report that evaluation and reporting of the ability of research data to meet the assumptions of parametric tests has been underemphasized despite the increased numbers of user-friendly computer packages available to test underlying test assumptions.

Types of Nonparametric Tests

A wide variety of nonparametric tests are available for use in health care research. After reviewing how to decide when to use nonparametric tests in Chapter 2 and evaluating the characteristics of data in Chapter 3, Chapters 4 through 9 will examine specific nonparametric tests that are available for use in most statistical computer packages. These nonparametric tests include those that evaluate "goodness of fit" (Chapter 4), matched samples and repeated measures (Chapter 5), repeated observations across multiple time periods (Chapter 6), differences between two or more independent groups (Chapters 7 and 8), and measures of association (Chapter 9). Chapter 10 will conclude with a summary table of the statistics reviewed in this text, along with an evaluation of areas in nonparametric statistics that are in need of further exploration.

2 The Process of Statistical Hypothesis Testing

A major function of both parametric and nonparametric statistics is statistical inference: We are interested in drawing conclusions about certain characteristics of a population of interest based on observations that we have obtained from a sample. To do this, we undertake the process of statistical hypothesis testing. To better understand this process, let us consider the hypothetical research example that was presented in the preface to this book and briefly examine the procedures that are common to most statistical hypothesis testing.

Recall that we are interested in the effects of a particular health care intervention on certain client outcomes. To be more specific, suppose that we have formulated the following research hypothesis:

> Children hospitalized for minor surgery who receive a specially designed clinical intervention will demonstrate more satisfactory postoperative outcomes with regard to cooperation, anxiety, fear, and posthospital adjustment than will hospitalized children who do not receive the intervention.

We propose to utilize a specific intervention with an experimental group of children ages 5 to 7 and compare those results with postoperative outcomes obtained from a control group of similarly aged children who will receive the customary patient care. Through a careful review of the research literature, we have also identified reportedly reliable and valid outcome measures to assess pre- and postintervention cooperation, anxiety, fear, and posthospital adjustment in this sample of young children. The design we propose is a classic pre- and posttest experimental design (Campbell & Stanley, 1966). A similar type of design and research hypothesis was utilized by Wolfer and Visintainer (1975) in their carefully

---

Box 2.1 Steps in Statistical Hypothesis Testing

Step Procedure to Be Followed

1. State the null hypothesis (H_0) and its research alternative (H_a).
2. Decide what data to collect and the conditions under which the data will be collected.
3. Specify the significance level (alpha) and determine whether alpha is one- or two-tailed.
4. Identify those statistical tests that would most satisfactorily answer the research questions formulated for the study.
5. Determine the desired sample size (N).
6. Collect the data, evaluate their properties, and select (from among the statistical tests identified) those that most satisfactorily meet the requirements for the study and whose assumptions are best met by the collected data.
7. If the research data meet the test's assumptions, compute the value of the test statistic. If the computed value is in the rejection region, reject H_0. If the value is outside the region of rejection, do not reject H_0.

---

planned clinical intervention study with hospitalized children funded by the U.S. Public Health Service (PHS).

To undertake the test of this research hypothesis, we would follow a seven-step approach (Box 2.1) similar to that outlined by Siegel and Castellan (1988). It should be emphasized that, although they are presented separately, each of these steps are interdependent. For example, availability or unavailability of certain types of data may alter the hypotheses that have been formulated and the statistical tests that are subsequently run.

1. State the null hypothesis (H_0) and its research alternative (H_a). The first step in any hypothesis testing procedure is that of stating the null and alternative hypotheses. These hypotheses are formulated and specifically stated for each dependent variable being examined. In our hypothetical example, one set of hypotheses could be stated as follows:

H_0: There are no differences between the intervention and control groups with regard to their posthospital adjustment.

H_a: The intervention group will demonstrate significantly more satisfactory posthospital adjustment than will the control group.

Note that the null hypothesis contains the statement of "no effect" and focuses on a single dependent variable. It should also be noted that although the alternative or research hypothesis usually is the true focus of clinical interest, it is the null hypothesis that will be tested in statistical analyses.

2. Decide what data to collect and the conditions under which the data will be collected. If we had sufficient funds, we might decide, as did Wolfer and Visintainer (1975), to collect data on 80 children who have been admitted to a particular hospital for minor surgery, randomly assign them to either the experimental or control group, collect pretest measures on our selected dependent variables, administer a carefully planned intervention, and measure postoperative outcomes on the same dependent variables. These outcomes might include level of pre- and postoperative anxiety and fear, degree of cooperation with the surgical procedures, and level of posthospital adjustment. Additional outcome measures could include pre- and postoperative pulse rate and time to first voiding (Wolfer & Visintainer, 1975).

In the present exploratory study, we propose to collect pre- and post-intervention information on a small sample of convenience of 20 children who have been admitted for minor surgery during a specified period of time to the hospital at which we are employed. The children, however, will be randomly assigned in equal numbers to the experimental and control groups.

3. Specify the significance level (alpha) and determine whether alpha is one- or two-tailed. The null hypothesis is the focus of statistical hypothesis testing. Based on the evidence that we have collected from our sample, we will decide to *reject* or *fail to reject* the null hypothesis. Because we are basing this decision on evidence obtained from a sample of observations and not an entire population, we can never be absolutely sure of the correctness of our decision.

There are two possible types of decision-making error that could occur. These are called *Type I* and *Type II* errors. Type I error occurs when, based on our sample evidence, we decide to reject the null hypothesis when, in fact, if we had evidence from the entire population, we would have ascertained that the null hypothesis was true. For example, we might conclude that, based on our sample evidence, there are differences between the intervention and control groups with regard to posthospital adjustment when in reality, the differences we observed were the result of error or random differences. Type II error occurs when, based on our sample evidence, we fail to reject the null hypothesis when, in fact, the null hypothesis is false. That is, we conclude that the evidence did not detect or demonstrate differences between the two groups when, in fact, differences exist.

The rates of occurrence of Type I and Type II errors are inversely related: Given the same sample size, decreasing the likelihood of one type of error will increase the likelihood of the other. In statistical hypothesis testing, the consequences of committing these errors need to be evaluated carefully given the context of the particular research. There are several excellent discussions presented in the statistics literature concerning both the setting of Type I and Type II errors and the use of confidence intervals as an alternative to hypothesis testing in statistical inference (e.g., Hays, 1994; Neter, Wasserman, & Whitmore, 1993).

In selecting the criteria for rejecting the null hypothesis, the researcher first needs to decide on the level of significance for his or her particular study. This level of significance, or alpha (α) level, represents the probability that the researcher will make the mistake of saying that differences exist between groups when in fact there are no differences.

Alpha must be set prior to conducting the study. Traditionally in health care research, alpha is set at .05, but as indicated, alpha could be set at any level depending on how serious it would be to make a Type I error. The statement "$\alpha = .05$" indicates that we are willing to make a Type I error 5 times out of 100.

Sometimes, because of the cost or potential negative side effects of the experimental intervention, we might want to be more certain that the observed differences between the groups are not just random error. We might, therefore, set our level of alpha at a more stringent level, for example, $\alpha = .01$. On the other hand, if the study is exploratory and the sample size is small, we might want to identify differences that, though not significant at $\alpha = .05$, might be of clinical interest. In such circumstances, we could set a more liberal alpha, for example, $\alpha = .10$. Remember, however, that if a more liberal alpha is set, the researcher increases the probability of committing a Type II error, or *beta error* (β).

Alpha provides us with guidance as to the conditions under which we will decide to accept or reject the null hypothesis. The larger alpha is, the greater the rejection region of the null hypothesis for the statistic we are considering. Thus, all other things being equal, $\alpha = .05$ will give us greater opportunity to reject the null hypothesis than will $\alpha = .01$.

A second criterion for establishing the rejection region for the null hypothesis is whether our research hypothesis is directional or nondirectional. For example, if the research hypothesis predicts that there will be a difference between two groups but the direction of difference is not stated, the region of rejection, or alpha, is equally divided between two possibilities. Group 1 could do better or worse than Group 2. If, on the other hand, the research hypothesis not only states that there is a difference between the groups but also states the direction of difference (e.g., Group 1 will do better than Group 2), then there is only one region of rejection for the null hypothesis, making alpha one-tailed.

In our hypothetical example, alpha is one-tailed because the research hypothesis states that the children in the intervention group will have more satisfactory outcomes than the children in the control group. By stating a direction, we are putting all of our $\alpha = .05$ "in one basket," thus increasing the chances of rejecting the null hypothesis—provided that the differences that we observe between the two groups are in the direction that we have predicted.

4. Identify those statistical tests that would most satisfactorily answer the research questions formulated for the study. The identification of a proposed plan for statistical analysis needs to take place prior to any data collection. The identified statistical tests will then provide a basis for determining an appropriate sample size and conducting a power analysis. A full determination of the appropriateness of a particular statistic, however, will be undertaken after the data have been collected.

5. Determine the desired sample size (N). Determining the sample size that is appropriate to a particular study is also undertaken prior to conducting the study. This is not an easy task and requires careful consideration of each of the research hypotheses posed and the statistical tests that have been selected for the proposed data analysis. For data that meet the assumptions of parametric tests, several textbooks are available to help the researcher to determine the most appropriate sample size (Cohen, 1988; Kraemer & Thiemann, 1987). Borenstein and Cohen (1988) also have published a very useful computer program, based on Cohen's (1988) classic text, that offers a hands-on, user-friendly approach to sample size and power analysis. Unfortunately, these approaches to sample size determination are based on analyses of parametric tests and are, therefore, less helpful when the research is exploratory in nature or when the distribution of the data is severely skewed, suggesting that nonparametric tests need to be considered.

Wolfer and Visintainer (1975) decided to set their one-tailed alpha at .05 and determined that a sample size of 80 children was sufficient to carry out their study at their desired level of power and effect. In our hypothetical exploratory study, we have determined that, based on financial considerations and necessity, at the maximum our sample will consist of 20 children. Moreover, our sample will not be randomly selected from all possible hospitals in our region but will be limited instead to all eligible children who enter a specific hospital during a stated period of time. We have, however, decided to randomly assign these 20 children in equal numbers to the experimental and control groups, thus maintaining the integrity of our quasi-experimental design.

6. Collect the data, evaluate their properties, and select from among the statistical tests identified those that most satisfactorily meet the requirements for the study and whose assumptions are best met by the collected data. We have collected the data on our two groups of children and are set to analyze the results. Now we need to determine which of the statistical tests that we identified in our proposed plan for data analysis would be most appropriate to use given our specific hypotheses and the characteristics of the collected data. We have identified several alternative tests, both parametric and nonparametric, that might be appropriate for our needs. How do we determine which test would be most appropriate for our situation? To make this choice, it is necessary to arrive at some criteria for choosing between a parametric and a nonparametric test.

Choosing Between a Parametric and a Nonparametric Test

There are both statistical and substantive criteria to be applied when choosing between parametric and nonparametric tests (Harwell, 1988). *Statistical* criteria refer to the test's ability to control the Type I error rate at user-specified alpha levels (e.g., .05) and the power of a particular test. *Substantive* criteria refer to nonstatistical criteria, particularly the level of measurement of the variables under consideration.

Controlling Type I Error Rate

Recall that Type I error rate is the extent to which we incorrectly state that there is a difference between the two groups when in reality there is not a difference. We have set this error rate prior to our collection of data (e.g., $\alpha = .05$). A "good" statistical test will control this alpha at the level we have specified. The ability of a test to control this Type I error, however, depends on the extent to which the data being analyzed meet the underlying assumptions of the test being considered. Certainly no data are perfect (one would be suspicious of data that are!), and certain violations of a test's assumptions are to be expected.

A second consideration in choosing a test, therefore, is the extent to which the test being considered is "robust" with respect to departures from its assumptions. Harwell (1988) cautions that overconfident researchers have tended to rely too heavily on the robust properties of parametric tests and have continued to use them even in the face of serious assumption violations. Monte Carlo evaluation procedures have indicated that, with a sufficient sample size (e.g., $N > 30$ per group), some tests such as ANOVA are fairly robust even in the face of substantial departures from normality and other assumption violations. Other tests, however, (e.g., ANCOVA, repeated-measures ANOVA, and multiple regression), are less able to withstand serious deviations from their assumptions.

Given the potential hazards of violating departures from normality, it is extremely important that the researcher consider the nature of the population from which the sample was drawn and whether it is realistic to expect that this population is, in truth, normally distributed. For example, would the pre-operation anxiety scores for our target population of all 5- to 7-year-old children hospitalized for minor surgery be normally distributed, or would it actually be negatively skewed, with a higher density of scores on the upper end of the anxiety scale?

Determining the Power of a Statistical Test

Given a parametric and a nonparametric test in competition, which test is more powerful? *Power* refers to the ability of a test to correctly reject the null hypothesis. A powerful test is also one whose assumptions have been sufficiently met. In comparing the power of two competing tests, therefore, the researcher needs to evaluate the ability of the data to meet the tests' assumptions. Other things being equal, when data sufficiently meet the assumptions of a parametric test, the parametric test generally is more powerful. When there are serious departures from the parametric test's assumptions, however, the nonparametric test generally is more powerful. Monte Carlo simulations have indicated that nonparametric tests tend to be more powerful than parametric tests when the distribution of the data being considered is unimodal but nonnormal in shape (Harwell, 1988).

Because the power of a statistical test increases as the sample size increases, it is possible to accommodate a less powerful test by increasing the sample size. The *power efficiency* of a statistical test refers to the increase in sample size that would be necessary to make one test (e.g., a nonparametric test) as powerful as its rival (e.g., a parametric test) given that the assumptions of the rival test have been met, the alpha level is held constant, and the sample size of the rival test, N_1, is also held constant (Siegel & Castellan, 1988). In this situation, the power efficiency of a test is determined as follows:

$$\text{power efficiency of Test } 1 = (N_2/N_1)100\%.$$

For example, if Test 1 requires a sample of 40 cases to have the same power as Test 2 with 30 cases, then Test 1 has a power efficiency of 75% ($[N_2/N_1]100\% = [30/40]100\% = 75\%$). A *power efficiency* of 75% implies that if the assumptions of both tests are met and Test 2 is more powerful than Test 1, then to achieve equality of power for the two tests, 10 cases would be required for Test 1 for every 7.5 cases used for Test 2 (Siegel & Castellan, 1988).

Level of Measurement of the Variables

There are additional substantive or nonstatistical criteria for selecting one test over another (Harwell, 1988). In particular, it is important to consider the level of measurement (i.e., nominal, ordinal, interval, or ratio) of the variables being considered for analysis. When the data being considered are nominal or categorical in measurement, there is little choice of tests to be undertaken; the test of choice is nonparametric. The issue becomes less clear when the data are ordinal, interval, or ratio. For readers who are unfamiliar with the levels of measurement of variables, Chapter 3 provides a review.

Traditionally, nonparametric tests have been intended for use with nominal or ordinal data, whereas parametric tests were used with dependent variables that were interval or ratio in measurement. The issues, however, are not only the variables' level of measurement but also the size of the sample and the shape of the parent distribution behind the data (Knapp,

---

Box 2.2 Situations That Suggest the Use of Nonparametric Tests

---

- The independent and/or dependent variables are nominal in measurement
- Ordered data with many ties
- Rank-ordered data specifying placement; no other metric assumed
- Unequal or small sample sizes
- Nonnormal distribution of the dependent variable
- Unequal variances across groups
- Unequal pairwise correlations across repeated measures
- Data with notable outliers

1990). Some of the most powerful parametric statistical analyses can occur with large sets of observations of ordinal data with normally shaped distributions and equal variances. Some of the least powerful parametric statistical analyses occur when the interval or ratio data have distributions that are skewed and the sample sizes are small.

The decision whether to use a parametric or nonparametric test, therefore, is not a simple one. The researcher needs to take into account not only the level of measurement of the variable(s) being considered but also the ability of the data being analyzed to meet the assumptions underlying the tests being considered. The most powerful test, parametric or nonparametric, is the one that best meets the underlying distribution of the data being considered in the hypotheses.

When to Consider a Nonparametric Test

Box 2.2 presents some situations that should suggest to the researcher that nonparametric tests are worthy of serious consideration. When both the dependent and independent variables consist of unordered categories (e.g., religious preference and marital status), no parametric test can be used. The researcher may also find that although the data may be orderable (e.g., two variables that have five categories of pain and level of discomfort), there are many ties in the data. This also suggests that nonparametric tests may be preferable to parametric tests. The researcher may also have rank-ordered data in which no other metric is assumed. Examples of this situation might be a student's placement on a test or in a competition (e.g., first, second, third).

Unequal or small sample sizes (e.g., 5 subjects per group) may prevent our being able to determine the shape of the sample data's distribution even when the population distribution is normal. In our hypothetical example, a sample size of only 10 children in each group suggests that it may be difficult to ascertain the data's underlying distribution. Even with larger sample sizes, the distribution of observations may give evidence of gross nonnormality (e.g., a heavily skewed, peaked, or flat distribution).

If several groups are being considered, there might also be unequal variances across groups, marked outliers, or unequal pairwise correlations across repeated measures. The bottom line is that the means are not truly representative of the scores of the sample; therefore, other measures of central tendency (e.g., the median) might be better descriptors of the scores.

Researchers can use a number of readily available approaches to evaluate the extent to which their data meet the assumptions of a particular test. Because of the importance of this topic to statistical hypothesis testing, Chapter 3 focuses on this topic.

7. If the research data meet the test's assumptions, compute the value of the test statistic. If the computed value is in the rejection region, reject H_0. If the value is outside the region of rejection, do not reject H_0. The final task of hypothesis testing is to determine whether or not to reject the null hypothesis. That is, assuming that our data meet the assumptions of our chosen test (and that is a *big* assumption), we would use the computer to determine the value of the test statistic and whether this calculated value is in our identified region of rejection of the null hypothesis. If it is, we would reject the null hypothesis and conclude that our experimental group did have more positive postoperative outcomes than the control group. If the calculated value of the test statistic is not in our identified region of rejection, we would fail to reject the null hypothesis and conclude that our evidence does not indicate any differences between the intervention and control groups with regard to postoperative outcomes. Again, for each decision, we are faced with possible error: concluding that there is a difference between the experimental and control groups with regard to postoperative outcomes when in reality there is no difference (Type I error), or failing to detect a difference between the two groups when in reality there is a difference (Type II error). A critical key to minimizing these error rates is our ability to choose the most appropriate test given the characteristics of our data. Chapter 3 presents some relatively easy approaches to the examination of these characteristics as a means of facilitating this choice.

3 Evaluating the Characteristics of Data

Chapter 2 focused on the process of statistical hypothesis testing. Part of this process (Step 6) involves evaluating the extent to which the data being analyzed meet the assumptions of the tests being considered. Chapter 3 will outline available methods for evaluating the characteristics of data. First, the level of measurement of a variable needs to be identified to determine the most appropriate parametric or nonparametric statistical test. Next, it is important to evaluate the normality of the variable's distribution, the impact of outliers, the homogeneity of variance, and sample size adequacy.

Characteristics of Levels of Measurement

Measurement is the process of assigning numbers or codes to observations according to certain prescribed rules. The way in which these values are assigned to the observations determines a variable's *level of measurement*. The most widely accepted set of rules for determining a variable's level of measurement is that developed by Stevens (1946). This typology consists of four levels of measurement whose order is based on how much information they carry. These levels are nominal, ordinal, interval, and ratio. Table 3.1 summarizes the characteristics of these four levels of measurement.

Nominal

The first level of measurement is nominal. A variable that is measured on a nominal scale is one that has distinct nonoverlapping categories. The

Table 3.1 Overview of the Characteristics of the Levels of Measurement

Level of Measurement	Mutually Exclusive Groups	Rank Ordering	Equidistant Values	Meaningful Zero Point	Example
Nominal	•				Marital status
Ordinal	•	•			Stress level (1–7)
Interval	•	•	•		Depression Scale (1–100)
Ratio	•	•	•	•	Weight (pounds)

numbers that are assigned to these categories have no intrinsic meaning, but all persons who share the same category are assigned a similar value.

There are three basic requirements for a "good" nominal-level variable: (1) All members of one level of the variable must be assigned the same number, (2) no two levels are assigned the same number, and (3) each observation can be assigned to one and only one of the available levels. Given that these three conditions have been fulfilled, the levels of the nominal-level variable are mutually exclusive and exhaustive.

The variable *gender* is a nominal-level measurement because it is composed of two independent, mutually exclusive (nonoverlapping), and exhaustive levels: male and female. In our hypothetical intervention study, each of the 20 participating children could be assigned a "0" or a "1" depending on whether the child is a male (0) or a female (1). The numbers 0 and 1 that have been assigned to these levels have no inherent order to them; these numbers could have been reversed. They merely indicate the gender group to which the child belongs. Additional variables in our hypothetical study that have a nominal level of measurement are the group to which the child was assigned: intervention (0) versus control (1), and type of minor surgery: tonsillectomy (1), appendectomy (2), ear surgery (3), or hernia repair (4).

Parametric statistics assume ordering and meaningful numerical distances between values; therefore, these statistics do not provide very useful information if the dependent or outcome variable has a nominal level of measurement. It does not make sense, for example, to report an average gender. For nominal data, researchers rely instead on frequencies, percentages, and modes to describe their results. Nonparametric inferential statistics (e.g., the chi-square goodness-of-fit test and the Fisher's exact test) may also be applied to these data.

Ordinal

The next level of measurement is ordinal. A variable that has an ordinal level of measurement is characterized by having mutually exclusive categories that are sorted and rank ordered on the basis of their standing relative to one another on a specific attribute according to some preset criteria. Although it may be possible to ascertain that one person has a higher rank relative to another person, it is not possible to determine exactly how much higher that person is than another.

Suppose the nurses in our hypothetical intervention study were asked to assess on a 7-point scale (1 = *not at all anxious* to 7 = *very anxious*) the extent to which a particular child is anxious prior to our planned intervention. This variable, preintervention anxiety, is an ordinal-level variable. We know, for example, that Child A, who received a "6" on preintervention anxiety, was rather anxious prior to treatment and was more anxious than Child B, who received a "3" on this scale. Because there are not equidistant intervals on this 7-point scale, however, it is not possible to conclude that Child A is twice as anxious as Child B or that the difference between a "6" and a "7" is the same as the difference between a "3" and a "4." Moreover, not all values necessarily share the same intensity. For example, Nurse C's assignment of a "7" to a child may not have the same intensity level as Nurse D's "7." We only know that, for both nurses, a particular child was "very anxious" according to their criteria.

Because there is order to the values of an ordinal scale, descriptive statistics that rely on rank ordering (e.g., the median) can be used in addition to percentages, frequencies, and modes. Numerous nonparametric inferential statistics are available to test hypotheses about similarities of medians between groups and relationships among variables.

There has been much heated discussion in the research literature about the appropriateness of using parametric tests with ordinal-level data (Armstrong, 1981; Burke, 1963; Gardner, 1975; Knapp, 1990). Pedhazur and Schmelkin (1991) suggest that this controversy was sparked by early writings of Stevens (1951), who argued that means and standard deviations, the backbones of parametric statistics, were not appropriate measures of central tendency for ordinal data. Others have effectively argued (Knapp, 1990) that the critical issue is not so much that the data are ordinal but rather that the data have a sufficient sample size (e.g., $N > 30$) and a relatively normal distribution of the dependent variable to merit the use of parametric statistics.

Interval

Interval-level scales are more refined than either nominal or ordinal scales. Like the ordinal scale, the interval-level scale has mutually exclusive groups and rank ordering. Unlike the ordinal scale, the interval-level scale has equidistant intervals. This means that we obtain information not only about the rank order of a particular score but also about how much greater or less a particular score is than another. That is, on an interval scale whose range is 1 to 100, the difference between 100 and 75 is, in some sense, the same as the difference between 75 and 50.

A classic example of an interval-level scale is temperature measured in degrees Fahrenheit. We know, for example, that a child whose body temperature is 102° has a temperature that is 2° higher than a child whose body temperature is 100°. Because an interval-level scale does not have an absolute zero point, however, the distances between values, although theoretically equidistant, do not carry exactly the same meaning. That is, the change in body temperature from 98° to 101° is not meaningfully the same as a change in body temperature from 102° to 105°. However, 100° is not twice as hot as 50° because 0° Fahrenheit is a numerical convenience, not an absolute.

A common practice among researchers is to use a multi-item scale to measure single or multiple constructs. The individual items tend to be either nominal (e.g., $0 = agree$ vs. $1 = disagree$) or ordinal (e.g., $1 = strongly$ $agree$ to $5 = strongly$ $disagree$) in nature, and the item responses are summed to produce a scale with interval-level properties and with a larger range of possible scores (e.g., 1 to 100). From these data, we can use all the measures of central tendency and variance. Parametric statistics such as the t test, ANOVA, and Pearson product-moment correlation coefficient are all possible considerations.

In our intervention example, we might decide to use a posthospital adjustment measure based on 30 items that parents completed. Similar to the scale used by Wolfer and Visintainer (1975), parents could be asked to rate on a scale of 1 to 4 ($1 = much\ less\ than\ before$ to $4 = much\ more\ than$ $before$) the extent to which their children expressed anxiety with regard to five areas related to their posthospitalization experience: sleep, eating, aggression, withdrawal, and separation. From these 30 items, a total posthospital adjustment score could be generated (range = 30 to 120) along with five 6-item subscale scores (range = 6 to 24) reflecting the five areas of adjustment.

Again, controversy exists as to the true nature of the level of measurement of such a multi-item scale (Gardner, 1975; Knapp, 1990; Nunnally, 1978; Nunnally & Bernstein (1994); Pedhazur & Schmelkin, 1991). That is, is an "interval" scale that has been generated from ordinal data truly interval? Should we even care? For statistical analysis, the concern is not so much the variable's "true" level of measurement as much as whether the information generated from the use of a particular statistic best represents the data. This conclusion can be reached only by examining the data thoroughly to determine the extent to which a particular test's assumptions have been violated. Pedhazur and Schmelkin (1991) indicate that, even in his later writings, Stevens (1968) argued, "The question is thereby made to turn, not on whether the measurement scale determines the choice of a statistical procedure, but on how and to what degree an inappropriate statistic may lead to a deviant conclusion" (p. 852).

Ratio

The highest level of measurement is ratio. In addition to maintaining the characteristics of the previous three levels of measurement (mutually exclusive and exhaustive categories, rank ordering, and equidistant intervals), a ratio-level variable also has a meaningful and absolute zero point that represents the complete absence of a given attribute. Because of its invariant zero point, the ratio of any two scores from a ratio scale is unchanged by transformations through multiplication and division.

Examples of ratio-level variables include weight, blood pressure, and temperature Kelvin. In our hypothetical study, a child's body weight and time to first voiding could be considered ratio-level variables. The age of the child might be more controversial. Our society has yet to agree on when an individual becomes a human being. At conception? At birth? Or at some other place along the way?

It does not matter much in statistics whether a variable is at the interval or ratio level of measurement. Both of these levels of measurement are appropriate for use with parametric statistics. To reiterate, equally important determinations regarding the use of parametric statistics are sample size and the shape of the distribution of the dependent variable.

Which Level of Measurement Is "Best"?

There is no clear answer as to which level of measurement is best for a particular research question. Clearly, the researcher wants to attain the very highest level of measurement possible given the time, financial, and design constraints of the research. The higher levels of measurement, interval and ratio, provide the researcher with the opportunity to use potentially more powerful statistical tests. Moreover, it is always possible to collapse data into lower levels of measurement. It is not possible, however, to resurrect interval-level data from precollapsed nominal data. The best approach is *not* to collapse data while entering them into the computer. Data can be collapsed, if necessary, later on during the statistical analyses.

Assessing the Normality of a Distribution

Returning to our hypothetical intervention study, suppose that we were interested in assessing the normality of the distribution of scores for children's preintervention anxiety. As indicated above, this is an ordinal-level variable whose scores can range from 1 to 7 (1 = *not at all anxious* to 7 = *very anxious*), with higher scores suggesting higher anxiety. There are several ways that we could assess the normality of this variable. First, we could examine the distribution's skewness and kurtosis. Next, we could visually examine the distribution of the data to obtain a sense of its shape. Finally, we could statistically test the extent to which the data fit a theoretically normal distribution.

All three of these approaches are available in SPSS for Windows by choosing the following commands from the menu: *Statistics . . . Summarize . . . Frequencies and Statistics . . . Summarize . . . Explore.* These commands open the *Frequencies* and *Explore* dialogue boxes and allow the researcher a number of options for evaluating data. By opening the *Frequencies . . . Charts* dialogue box and selecting *Histograms . . . with normal curve*, a normal distribution is superimposed over the histogram of the variable of interest. This allows the researcher to visually inspect the data for violations of normality. The *Statistics . . . Summarize . . . Explore* command may also be used to statistically test for normality. This procedure also produces information regarding descriptive statistics, stem-and-leaf plots, boxplots, descriptions of outliers, normal probability plots, and statistical tests of normality. Separate analyses can be obtained for subgroups of data as well.

Skewness

You will recall that a normal distribution takes the form of a bell-shaped curve that is centered on the mean (Figure 3.1A). The normal distribution is symmetric and all three measures of central tendency—the mean, median, and mode—share the same value. One simple way of assessing normality of a distribution, therefore, is to examine the measures of central tendency. If the mean, median, and mode are nearly equal in value, then there is evidence to suggest that the distribution is symmetric. If these three values are not at all similar, then the distribution is characterized as being asymmetric or *skewed*; that is, the distribution has one tail that is longer than the other.

There are two types of skewness: positive and negative skew. A distribution is *positively skewed* if the distribution's longer tail extends toward the right or toward the higher set of values (Figure 3.1B). This results in the value of the mean being pulled to the right and being larger in value than the median or mode. A distribution that is *negatively skewed* has a longer tail that extends toward the left or toward the lower set of values (Figure 3.1C). This results in the mean being smaller in value than the median or mode.

The measure of skewness is referred to as the third moment about the mean, $\Sigma(X - \overline{X})^3/(n - 1)$ (Neter, Wasserman, & Whitmore, 1993). Because this third moment is measured in cubed units (e.g., weight cubed), a standardized measure of skewness is considered to be more useful because its size does not depend on the units of measurement. This standardized measure is obtained by dividing the third moment by the cube of the standard deviation (s^3) of the variable being examined (Neter et al., 1993). This is the skewness value that is presented in the computer printout. When a distribution is a symmetric bell-shaped curve, the value of this measure of skewness is 0. The measure has a negative value when the distribution is negatively skewed and a positive value when the distribution is positively skewed. To determine the seriousness of the skewness of a distribution, one of two measures of skewness, Fisher's or Pearson's, can be used (Hildebrand, 1986; Lehman, 1991; Munro & Page, 1993). The Fisher's coefficient is as follows:

Fisher skewness coefficient = skewness/standard error skewness.

To calculate the Fisher skewness coefficient, the computer-generated value for skewness (Skewness) is divided by the standard error for skewness

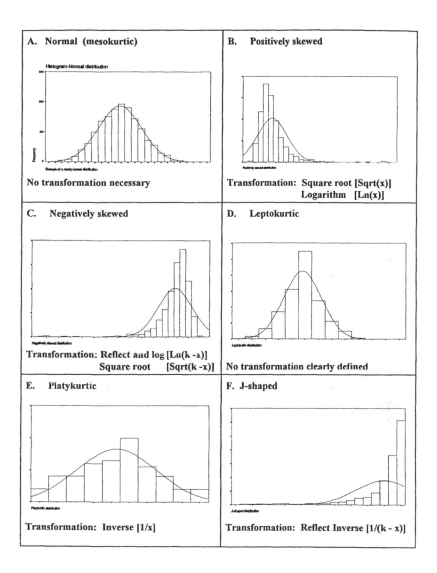

Figure 3.1. Comparison of the Most Common Forms of Distributions and Suggested Transformations

NOTE: Transformation suggestions come from Hair, Anderson, Tatham, and Black (1995) and Tabachnick and Fidell (1989). In Panels C and F, k represents a constant, usually the largest score + 1.

(S E Skew). If the resulting z statistic lies beyond the range of ± 1.96 (the critical value for a two-tailed z statistic at $\alpha = .05$), the distribution is asymmetric and significantly skewed. Calculated values of this coefficient that fall between -1.96 and $+1.96$ suggest that the distribution is not significantly different from a normal distribution.

A second commonly used index for skewness is the Pearson skewness coefficient (Sk_p):

$$\text{Pearson skewness coefficient} = Sk_p = 3[(\overline{X} - Md)/s].$$

This statistic uses the difference between the mean, $(\overline{X})$, and median (Md) of a distribution divided by the variable's standard deviation (s) to determine the level of skewness. If $Sk_p = 0$, the mean and median are equal and therefore the distribution is symmetric. A negative coefficient indicates a negative skew (i.e., the mean is smaller than the median), and a positive value represents positive skewness (i.e., the mean is larger than the median). Lehman (1991) suggests that values of Sk_p between -0.5 and $+0.5$ indicate generally acceptable levels of skewness.

Kurtosis

A second characteristic of a distribution is its *kurtosis*, or the fourth moment about the mean (Balanda & MacGillivray, 1988; Neter et al., 1993), calculated as $\Sigma(X - \overline{X})^4 /(n - 1)$. Because this fourth moment is measured in units^4, a standardized measurement of skewness is available in statistical packages that divides the fourth movement by s^4. Evaluation of a distribution's kurtosis is especially useful after it has been determined that the distribution is not unduly skewed. It is not very useful for asymmetric or skewed distributions.

Three terms are used to denote different levels of kurtosis: *mesokurtic*, *leptokurtic*, and *platykurtic*. A normal distribution has a standardized kurtosis value that is equal to zero and is referred to as being *mesokurtic* (Figure 3.1A). A positive value for the standardized kurtosis coefficient implies that the distribution is *leptokurtic*, or more peaked than a normal distribution (Figure 3.1D). (To remember what *leptokurtic* means, it might be helpful to recall Superman, *leaping* tall buildings in a single bound.) A negative value for the standardized kurtosis coefficient implies that the distribution is *platykurtic*, or flatter than a normal distribution (Figure 3.1E). (Remember that, like the platykurtic distribution, a platypus is an animal that stands low and close to the ground.)

Munro and Page (1993) suggest using a Fisher coefficient to evaluate kurtosis:

Fisher coefficient of kurtosis = kurtosis/standard error of kurtosis.

That is, the standardized kurtosis value is divided by its standard error to determine the extent to which a bell-shaped symmetric curve deviates from a normal distribution. If this z statistic falls outside the range of ±1.96, then the bell-shaped distribution is significantly different from a standard normal distribution.

Computer Analysis of Skewness and Kurtosis

Assuming that the data for the 20 subjects in our hypothetical study have been entered into the computer data file, the SPSS for Windows syntax commands and computer-generated frequency output for the variable Pre-intervention Anxiety (Anxiety1) are presented in Table 3.2. This output was obtained by clicking on the commands *Statistics . . . Summarize . . . Frequencies* and selecting the Anxiety1 variable for analysis.

We are first presented with the frequency distribution of the preintervention anxiety variable, Anxiety1 (①). Given that no child indicated that he or she was "not very anxious" during the preintervention phase, there are no values given for "1" or "2." Notice that the mean, median, and mode are not equal to one another, suggesting that the data may be skewed (②). Because the mean is smaller than either the median or the mode, the data are negatively skewed, with the longer tail in the direction of the smaller values, a condition that is verified by the skewness value of −1.103 (③). Dividing the measure of skewness by the standard error for skewness (−1.103/.512) results in a Fisher skewness coefficient of −2.15, which falls outside the acceptable limits of ±1.96, suggesting that the data may be seriously skewed. It is interesting that a different result is obtained for the Pearson Sk_p:

$$Sk_p = 3(\overline{X} - Md)/s = 3(5.85 - 6.0)/1.309) = -.34.$$

The resulting value of −.34 *is* within the acceptable range of this coefficient (−.5 to +.5). This discrepancy may be explained by the extreme sensitivity of the Fisher measure of skewness to outliers (Munro & Page, 1993). Because the statistic is based on deviations from the mean raised to the third power, outliers have a very strong effect on this measure.

Table 3.2 Syntax Commands and Computer Printout of Frequencies for Pretreatment Anxiety

```
FREQUENCIES
   VARIABLES=anxiety1
   /STATISTICS=STDDEV VARIANCE RANGE MINIMUM MAXIMUM SEMEAN MEAN MEDIAN MODE
   SUM SKEWNESS SESKW KURTOSIS SEKURT
   /HISTOGRAM  NORMAL.

ANXIETY1  pretreatment anxiety
①                                                          Valid       Cum
                              3.00        2      10.0      10.0      10.0
                              4.00        1       5.0       5.0      15.0
                              5.00        3      15.0      15.0      30.0
                              6.00        6      30.0      30.0      60.0
very anxious                  7.00        8      40.0      40.0     100.0
                                       -------   -------   -------
                             Total      20     100.0     100.0

②
Mean          5.850    Std err       .293    Median         6.000
Mode          7.000    Std dev      1.309    Variance       1.713
Kurtosis       .380 ④  S E Kurt      .992    Skewness      -1.103 ③
S E Skew       .512    Range        4.000    Minimum        3.000
Maximum       7.000    Sum        117.000
Valid cases     20     Missing cases      0
```

SOURCE: Generated from SPSS for Windows.

Ordinarily, when a distribution has serious skewness problems, indicating that it is not bell-shaped, it would not be necessary to examine its kurtosis. Given that we have conflicting information regarding this distribution's skewness, however, it would be useful to examine the distribution's kurtosis as well. The positive value (.380) for kurtosis (④) indicates that the distribution is leptokurtic, or more peaked than a normal distribution. Dividing the value for kurtosis by its standard error (.992), however, yields a Fisher coefficient of kurtosis of .383, which is well within the ±1.96 range for a normal distribution.

Visually Examining the Shape of the Distribution

Given these somewhat conflicting results, it is important that we examine the data visually to determine for ourselves the seriousness of the skewness. In fact, the necessity of visually examining data for departures from normality cannot be overstressed. No statistical test of normality is superior to what my biostatistician friend, Dr. James Reading, refers to as the *eyeball test*.

The eyeball test consists of visually examining the data's distribution to determine if the distribution looks sufficiently comparable to a normal distribution for the researcher to feel comfortable using parametric tests. Is the mean an adequate representation of these data? Are there unusual "kinks" in the distribution? Is the distribution unimodal, or are there multiple modes? Are there outliers about which to be concerned? What effect does the sample size have on the potential shape of the distribution? Although the mean, median, and mode may be similar, a limited sample size may restrict one's ability to adequately distinguish the shape of the distribution. If the data do not have a normal distribution, is there a possible transformation that could be performed (e.g., log or square root) that would make sense logically and that would transform the nonnormal distribution into a more nearly normal distribution?

Figure 3.2A presents a graph of the normal curve superimposed on the preintervention anxiety (Anxiety1) distribution for the 20 subjects in our hypothetical study. This figure indicates that the data are negatively skewed and somewhat leptokurtic in shape. The distribution also appears to have serious deviations from normality. With so few data points ($N = 20$), the shape of the distribution may also not be definitively determined. This information suggests that the use of nonparametric tests with these data may be in order. A second alternative would be to consider the possibility of transforming the preintervention anxiety variable to obtain a more nearly normal distribution.

There are additional plots of normality that may be generated in SPSS for Windows through the *Statistics . . . Summarize . . . Explore* command that can be of help in visually examining data. These plots include normal probability plots and detrended normal plots.

Normal and Detrended Normal
Probability Plots

Figure 3.3 presents examples of normal and detrended normal probability plots for selected distributions. In the *normal probability plot*, each data point is paired with its expected value given a nearly normal distribution of similar range and sample size. If the sample is from a nearly normal distribution (Figure 3.3A), a normal probability plot of the observed and expected values would indicate that nearly all values lie along a 45° straight line running from the lower left corner to the upper right corner of the

text continued on p. 44

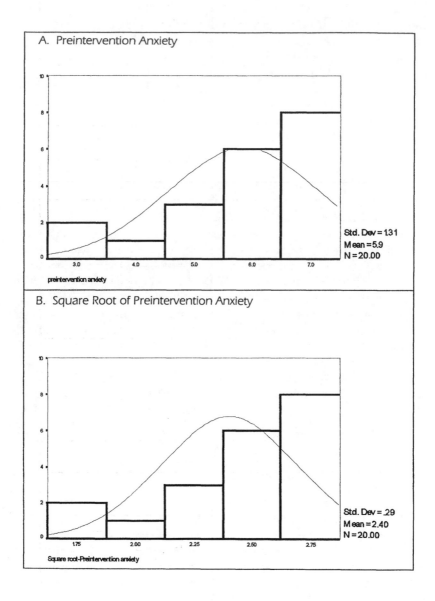

Figure 3.2. Histograms of Preintervention Anxiety (Anxiety1) and Its Square Root Generated From SPSS for Windows

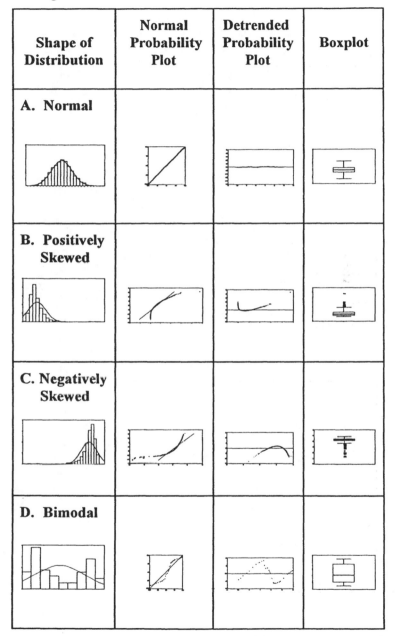

Shape of Distribution	Normal Probability Plot	Detrended Probability Plot	Boxplot
A. Normal			
B. Positively Skewed			
C. Negatively Skewed			
D. Bimodal			

Figure 3.3. Examples of Normal Probability, Detrended Normal Probability, and Boxplots for the Normal and Other Selected Distributions

plot. A normal probability plot for a normal distribution is presented in Figure 3.3A, in the second panel. Note that, except for a few minor deviations, the values fall along the 45° line.

A *detrended normal probability plot* is one in which the deviations from normal for each value in the sample are plotted against the observed values. If the sample is from a nearly normal distribution, these deviations will cluster evenly around zero along a horizontal band. This indicates that there is little difference between the observed values and expected values. The detrended normal probability plot for the nearly normal distribution in Figure 3.3A, third panel, illustrates this pattern. Note that the data do *not* need to fall *exactly* along a straight line but rather that the band of values is similar in width across all values of the data.

Distributions that are skewed or bimodal (e.g., Figures 3.3 B-D) show markedly different patterns of deviations from normality. Curvilinear patterns often emerge, suggesting that either the data are badly skewed (Figures 3.3 B-C) or bimodal (Figure 3.3D). Outliers can be identified on these plots because they occupy positions away from the other values and do not appear to be connected to them (Tabachnick & Fidell, 1996). For example, in Figure 3.3C, the values in the lower left-hand corner of the normal and detrended normal probability plots represent the outliers for this negatively skewed distribution.

Computer Examples of the Plots

Figure 3.4 presents normal probability and detrended normal probability plots for the preintervention anxiety variable (Anxiety1). The plots were generated from the *Statistics . . . Summarize . . . Explore* commands for SPSS for Windows.

The two plots in Figure 3.4 indicate that the preintervention anxiety data are not normally distributed. The values for Anxiety1 are not similar to the expected values and, therefore, are not situated on the 45° straight line of the normal probability plot (Figure 3.4A). The detrended plot (Figure 3.4B) indicates that the largest deviation from normality appears to be with the smaller values; they are farthest both above and below the horizontal line that goes through 0.

Statistical Tests of Normality

The statistical tests for normality that are provided in SPSS for Windows are the Shapiro-Wilks and Kolmogorov-Smirnov (K-S) Lilliefors statistics.

A.   Normal Probability Plot

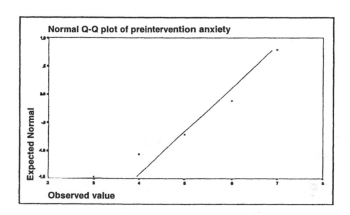

B.   Detrended normal probability plot

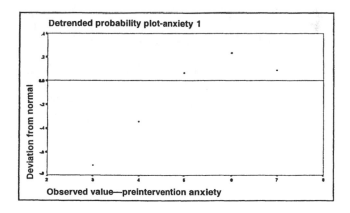

Figure 3.4. Normal Probability Plots of Pretreatment Anxiety Scores
(*N* = 20)

Table 3.3 Statistical Tests for Normality of the Preintervention Anxiety
Variable

	Statistic	df	Significance
Shapiro-Wilks	.8106	20	< .0100 ①
K-S (Lilliefors)	.1898	20	.0572 ②

SOURCE: Generated by SPSS for Windows.

These can be obtained by selecting the *Statistics . . . Summarize . . . Explore* commands from the menu, clicking on *Plots* and selecting *Normality plots with tests*. The objectives of these nonparametric goodness-of-fit tests are to compare the obtained distribution with a theoretically normal distribution and to determine whether the deviations from normality are sufficiently large to conclude that the distribution under investigation is not normal. The null hypothesis is that the data are normally distributed; the alternative hypothesis is that the data are not normally distributed. The null hypothesis will be rejected if the obtained significance level is less than our stated level of alpha (e.g., $\alpha = .05$).

Both the Shapiro-Wilks and K-S Lilliefors statistics are extremely sensitive to departures from normality. It is strongly recommended, therefore, that the researcher supplement these statistical tests with the previously discussed methods for examining data for departures from normality (e.g., visually examining the data and assessing skewness and kurtosis).

The computer printout generated from SPSS for Windows for the Shapiro-Wilks and K-S Lilliefors statistics is presented in Table 3.3. We seem to be obtaining conflicting results from the Shapiro-Wilks and K-S Lilliefors statistics. The Shapiro-Wilks test indicates that the distribution is not normal (significance < .01 is less than $\alpha = .05$; see Table 3.3, ①), whereas the K-S Lilliefors statistic suggests that the data are within our criteria of retaining the null hypothesis of normality (.0572 > .05, ②). Conover (1980) suggests that the Shapiro-Wilks test for normality may be more powerful than the K-S Lilliefors statistic in that it may be more likely to correctly reject the null hypothesis of normality. Because, in this situation, we would prefer not to reject the null hypothesis and accept our data as a normal distribution, which of these conflicting results should we believe?

Our determination of whether to accept or reject the preintervention anxiety distribution as normal should be based on all contributing factors: the level of measurement of the data, its visual representation, the similarity of the measures of central tendency, skewness and kurtosis, the statistics, and the sample size. Based on this evidence, we would most likely conclude that the data for preintervention anxiety are *not* normally distributed. This conclusion is based on the observation that these are ordinal data; the visual representations suggest nonnormality; the mean, median, and mode are not similar; there is some skewness; the Shapiro-Wilks and K-S Lilliefors statistics are close enough to our level of alpha to support rejection of the null hypothesis of normality; and we had a sample size of only 20. This determination would suggest that we would seriously need to consider using nonparametric statistics when analyzing this variable.

Examining Distributions of the Dependent Variable by Subgroups

For many parametric tests, it is expected that the distribution of the dependent variable be normally distributed not only as a whole but also when broken down into subgroups of a particular independent variable of interest. Table 3.4 presents the syntax commands and a breakdown of the preintervention anxiety scores of the children by intervention and control groups. These printouts were generated in SPSS for Windows by highlighting the *Statistics . . . Summarize . . . Explore* commands in the menu and placing the dependent variable, Anxiety1, in the *Dependent List* and the independent variable, Group, in the *Factor List.*

The resulting descriptive statistics and stem-and-leaf plots (Table 3.4, ①) indicate that the experimental and control groups have similar means and distributions. This suggests that we may have been successful in creating similar groups through random assignment—at least with regard to preintervention anxiety. The skewness statistics for the intervention group (skewness/standard error for skewness = $-1.0848/.6870 = -1.579$) and the control group ($-1.3379/.6870 = -1.95$) also indicate that the skewness is within acceptable range (± 1.96). Given the small sample size for both groups ($n = 10$), however, as well as the shape of the stem-and-leaf plots for both groups and the level of measurement of the variable (ordinal), nonparametric tests most likely would be used with these data. This conclusion is further supported by the significant Shapiro-Wilks tests (②) for both groups (.0458 and .0344 are less than our $\alpha = .05$).

Table 3.4 Pretreatment Anxiety by Group (intervention, control)

```
EXAMINE
  VARIABLES=anxiety1 BY group
  /PLOT BOXPLOT STEMLEAF NPPLOT
  /COMPARE GROUP
  /STATISTICS DESCRIPTIVES
  /CINTERVAL 95
  /MISSING LISTWISE
  /NOTOTAL.

pretreatment anxiety
By  GROUP              .00  intervention

Valid cases: 10.0  Missing cases:   .0  Percent missing:   .0
Mean      5.8000   Std Err   .4422   Min   3.0000  Skewness   -1.0848
Median    6.0000   Variance 1.9556 ③ Max   7.0000  S E Skew     .6870
5% Trim   5.8889   Std Dev  1.3984   Range 4.0000  Kurtosis     .2648
95% CI for Mean  (4.7996, 6.8004)    IQR   2.2500  S E Kurt    1.3342

Frequency    Stem &  Leaf ①
    1.00       3 .  0
    1.00       4 .  0
    1.00       5 .  0
    3.00       6 .  000
    4.00       7 .  0000
                       Statistic     df      Significance ②
Shapiro-Wilks            .8369       10          .0458
K-S (Lilliefors)         .1954       10         > .2000
-------------------------------------------------------------------
    ANXIETY1  pretest anxiety

By  GROUP            1.00  control

Valid cases: 0.0  Missing cases:   .0  Percent missing:    .0
Mean      5.9000   Std Err   .4069   Min   3.0000  Skewness   -1.3379
Median    6.0000   Variance 1.6556 ④ Max   7.0000  S E Skew     .6870
5% Trim   6.0000   Std Dev  1.2867   Range 4.0000  Kurtosis   1.8636
95% CI for Mean  (4.9796, 6.8204)    IQR   2.0000  S E Kurt    1.3342

Frequency    Stem &  Leaf ①
    1.00       3 .  0
     .00       4 .
    2.00       5 .  00
    3.00       6 .  000
    4.00       7 .  0000

                       Statistic     df      Significance ②
Shapiro-Wilks            .8245       10          .0354
K-S (Lilliefors)         .1963       10         > .2000
-------------------------------------------------------------------
Test of homgeneity of variance       df1     df2  Significance

    Levene Statistic         .1470     1      18      .7059 ⑤
```

SOURCE: Printout generated from SPSS for Windows.

Dealing With Outliers

One of the disadvantages of the mean as a measure of central tendency is its sensitivity to outliers. Because outliers are extreme data points that are very much different from the rest of the data, they tend to pull the value of the mean in their direction. This can result in serious distortion of results. The median, on the other hand, is not at all influenced by atypical data points because the median assesses ranks, not actual values. The presence of outliers, therefore, requires a careful assessment of their influences both on the mean and on the variable's distribution. Outliers also provide information about the types of cases that may not fit a particular hypothesized model.

There are two types of outliers: univariate and multivariate. Univariate outliers are those cases that possess extreme values on a single variable (e.g., a child who has an extreme stress score). Multivariate outliers are cases with unusual combinations of scores on two or more variables. For example, a person may be of an acceptable age (e.g., 16 years old) and another person could have a reasonable number of children (e.g., four), but a 16-year-old who has four children would most likely appear as a multivariate outlier.

Assessing Univariate Outliers Using the Boxplot

Boxplots (Figure 3.3) are very useful for identifying cases that are univariate outliers. They also provide a snapshot summary of the descriptive statistics for the distribution. On request, SPSS for Windows plots the smallest and largest values of the data set, the median (the horizontal bar inside the box), the 25th percentile (the lower boundary of the box), and the 75th percentile (the upper boundary), and it presents values that lie far outside this range. The interquartile range makes up the box presented in this plot. This is where 50% of the cases are located. The boxplot for the normal distribution in Figure 3.3A illustrates a distribution that is symmetrical, with equal tails, and a median that lies halfway between the upper and lower boundaries of the box.

Two types of univariate outliers are presented in the boxplots for SPSS for Windows (Norusis, 1995a). Any value that is more than 3 box-lengths (i.e., $3[P_{75} - P_{25}]$) from the upper or lower boundary of the box is designated on the plot with a "*" and is referred to as an *extreme value*. Each value that is between 1.5(i.e., $1.5[P_{75} - P_{25}]$) and 3 box-lengths from the upper or lower boundary of the box is identified with an "O" and is called an *outlier*. The outliers and extreme values are also identified either by their case number (the default option) or by specifying a case label (e.g., the variable *id*). This information is useful for tracking down and correcting possible errors in data entry. The largest and smallest observed values that are not outliers are presented by lines drawn from the ends of the box to these values.

In general, boxplots are useful for comparing the distribution of a continuous variable for two or more subgroups in a sample. For example, Figure 3.3, in panels B-D, presents the boxplots for a positively skewed, a negatively skewed, and a bimodal distribution. The boxplots for the positively and negatively skewed distributions indicate that the distributions are asymmetrical, having a long tail in one direction. The median in each case is no longer in the middle of the box but rather lies closer to the bottom or top of the box, depending on the type of skew. Extreme values (*) and outliers (O) can also be found lying beyond the longer tail. It is interesting that the boxplot for a bimodal distribution (Figure 3.3D) is not very helpful in revealing the shape of the distribution. Although the box for this distribution is very large compared to the tails and there are no outliers, its bimodal shape has become hidden.

Boxplots are especially useful for comparing two distributions. For example, the boxplots for the preintervention anxiety scores for the intervention and control groups are presented in Figure 3.5. These boxplots confirm our suspicion, based on visual inspection, that the preintervention anxiety data are negatively skewed for both groups: There is only one tail presented, directed toward the lower end of the values. Had the data been more normally distributed, two tails of equal length would have been presented, and the boxplots would have been similar to that in Figure 3.3A.

The lack of an upper tail for the preintervention anxiety scores is understandable because there is a restricted range for this variable (1-7). The 75th percentile for this distribution is identified in the graph as the value 7 and the 25th percentile as the value 5. Because three box-lengths is equal to 6 ($3[P_{75} - P_{25}] = 3[7 - 5] = 6$) and 1.5 box-lengths is equal to 3 ($1.5[P_{75} - P_{25}] = 1.5[7 - 5] = 3$), the extreme values (*) for this example would be those values that are either 13 or larger (7 + 6 = 13) or −1 or smaller. Outliers (O) would be 1.5 box-lengths above and below the upper and lower boundaries of the box, or the values of 10 (7 + 3 = 10) and 2 (5 − 3 = 2), respectively. Note that none of these extreme values is possible for this ordinal-level variable because the range of possible values for this Likert-type scale was 1-7. No children reported scores of less than 3, and

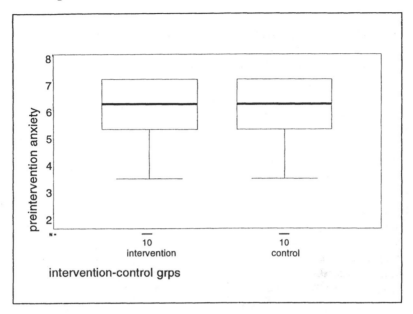

Figure 3.5. Boxplots of Preintervention Anxiety for Intervention and Control Groups (generated in SPSS for Windows)

a score of 10 is outside the possible range, so there were no outliers. Because there were no extreme values or minor outliers for this distribution, there are no "*" or "O" symbols in the computer printout. The conclusion to be drawn, therefore, is that the distribution of these data for both groups is very compact, of low range, and not normal.

Assessing Multivariate Outliers

Although the boxplot provides useful information about univariate outliers, it does not tell us anything about cases that have unusual patterns of scores with respect to two or more variables. These multivariate outliers can be screened by computer using techniques made available within SPSS using its regression analyses. Because the focus of this text is on nonparametric statistics, we will not examine these issues here. For the interested reader, these techniques (e.g., examining linear relationships, use of the Mahalanobis distance, and approaches to the analyses of residuals) are described in great detail and clarity by Tabachnick and Fidell (1996) and

Hair, Anderson, Tatham, and Black (1995) in their excellent textbooks on multivariate statistical analysis.

What to Do About Outliers

Researchers appear to have mixed feelings about outliers and what to do about them. Some researchers view outliers as nuisance cases, ones that do not fit expectations. Others suggest that the outliers in a study are the cases that should be examined most closely. Kruskal (1988), for example, argues that, "Miracles are the extreme outliers of nonscientific life. . . . It is widely argued of outliers that investigation of the mechanism for outlying may be far more important than the original study that led to the outlier" (p. 929).

A critical task for the researcher is to determine why outliers exist in the first place. Are they a result of errors of coding or measurement, or are they legitimate cases that possess unique characteristics with respect to one or more variables? Different approaches to remedying problematic outliers and reducing their influence have been suggested, depending on the etiology of the outlier's presence (Hair et al., 1995; Johnson, 1985; Pedhazur & Schmelkin, 1991; Tabachnick & Fidell, 1996). Such techniques include eliminating the case altogether, reweighting or recoding the outlier to reduce its influence, and transforming the variable to create a more nearly normal distribution. It may also be useful to analyze the data both with and without the extreme data points to determine the extent of the outliers' influence.

An enormous advantage of nonparametric rank-order statistics is that the ranking of data that occurs with these statistics serves to reduce the influence of outliers because the data being analyzed are ranks, not actual scores. There is no "quick fix" to the problem of outliers, and careful attention must be paid to the consequences of a particular remedy. These decisions must also be duly reported in the data analyses.

Data Transformation Considerations

When a particular distribution of a variable does not meet the normality assumption, it is possible to transform the values of that variable to create a new variable that has a more nearly normal distribution. Although this process appears easily accomplished, it does have serious problems, particularly with regard to both finding an adequate transformation index that will produce a more nearly normal distribution and interpreting the results of such a transformation. Figure 3.1 presents several common forms of nonnormal distributions and some suggested transformations that might help to create a more nearly normal distribution for the transformed variable. Hair et al. (1995) suggest that for flat (platykurtic) distributions (Figure 3.1E), the most common transformation is the inverse ($1/x$). A variable that is positively skewed (Figure 3.1B) might benefit from a log transformation ($\log(x)$), whereas one that is negatively skewed (Figure 3.1C) might be altered with a square root transformation. Leptokurtic distributions (Figure 3.1D) do not appear to have clearly defined transformations available in the research literature. Hair and colleagues (1995) also indicate that to achieve a noticeable effect from a transformation, the ratio of a variable's mean to its standard deviation should be less than 4.0 (i.e., mean/standard deviation < 4.0).

The goal of transforming data is to obtain a new distribution that is nearly normal in shape, with few outliers, and with skewness and kurtosis values near zero (Tabachnick & Fidell, 1996). It is important, therefore, that the researcher closely examine the distribution of the resulting transformation to ascertain if this goal has been achieved. Next, a careful interpretation of the resulting transformation needs to be made. Remember that a transformed variable no longer carries the original interpretation; the square root of preintervention anxiety is *not* the same as preintervention anxiety. Interpreting the meaning of a transformed variable is one of the most challenging tasks for the researcher.

In an attempt to obtain a more nearly normal distribution, the preintervention anxiety variable, Anxiety1, was transformed using the square root transformation indicated for negatively skewed distributions. Undertaking transformations of variables can be undertaken easily in SPSS for Windows through its *Transform . . . Compute* command. A new target variable, Sqrtanx1, was obtained by indicating that it represents the square root of the old variable, Anxiety1 (Sqrt(Anxiety1)). The syntax command in SPSS for Windows for this transformation is as follows: Compute Sqrtanx1 = Sqrt (Anxiety1).

Figure 3.2 compares the newly formed square root transformation with the original preintervention anxiety distribution. If the goal of data transformation is to obtain a nearly normal distribution with few outliers and with values of skewness and kurtosis near zero, it is apparent that this transformation has not been successful. This lack of success may result from the limited scale values (1-7). It may also be a result of our having not met the criteria set by Hair and colleagues (1995) of mean/standard deviation < 4; here, $5.85/1.309 = 4.47$. This lack of success suggests that nonparametric statistics, which rely predominantly on the ranking of data, may be the approach of choice.

Examining Homogeneity of Variance

Another important assumption of parametric tests that compare differences between two or more groups is that the variances among the subgroups must be similar; that is, there is homogeneity of variance. A general rule of thumb is that the variance of one group should not be more than twice that of another. This assumption is especially important when groups of unequal size are being compared (Tabachnick & Fidell, 1996).

Several tests of homogeneity of variance are available in SPSS. These include Box's M and the Levene test. The *SPSS for Windows base system user's guide (Release 6.0)* (Norusis, 1995a) suggests that the Levene test is less dependent on the assumption of normality than are most tests of homogeneity of variance. The null hypothesis for all tests of homogeneity is that the variances among the groups are equal, whereas the alternative hypothesis states that the variances are unequal. The null hypothesis will be rejected if the obtained level of significance is less than the preset level of alpha (e.g., $\alpha = .05$).

The descriptive statistics presented for the preintervention anxiety variable (Anxiety1) in Table 3.4 indicate that the variance for the experimental group was 1.9556 (③), which can be compared to 1.6556 for the control group (④). Because one variance is less than twice the other, it would appear that the homogeneity of variance assumption for preintervention anxiety has been met. The resulting Levene test (⑤) generated from the *Statistics . . . Summarize . . . Explore* command indicates that we would indeed fail to reject the null hypothesis of equal variances because the significance level (.7059) is considerably greater than our $\alpha = .05$. We should be pleased with this "failure" because we can conclude that the variances between the groups are equal.

Evaluating Sample Sizes

Two additional issues that are important determinants of which statistic—parametric or nonparametric—is most appropriate to use in an analysis are the absolute size of the sample being used and whether there are equal numbers of subjects in the subgroups being analyzed.

Minimum Sample Size Requirements

Determining an appropriate sample size for a study is a challenging and, at times, formidable task. As Cohen (1988) and Kraemer and Thiemann (1987) have indicated, sample size consideration is directly related to statistical power, or the researcher's ability to correctly reject a null hypothesis. All other factors being equal (e.g., similar effect size and alpha level), the larger the sample size, the more likely it is that the researcher will be able to detect significant differences among groups. Smaller sample sizes reduce statistical power and increase the researcher's chance of making a Type II error, that is, incorrectly concluding that the null hypothesis is true.

There does not appear to be any definitive agreement among statisticians as to how large a *large* sample is or what size sample is *too small* for use with a particular parametric test. An examination of the statistics literature indicates that the process of determining "how large is large enough?" is inextricably linked to issues of statistical power and also needs to be confined to the requirements of specific statistical tests (e.g., multiple regression, *t* tests, and ANOVA). Wampold and Drew (1990) argue that the answer to the question of sample size also depends in part on the normality of the distribution of the dependent variable of interest. They point out that whatever the form of the original distribution, the Central Limit Theorem has shown that as the sample size increases, the sampling distribution of the *mean* approaches normality. Because the mean plays a major role in most parametric statistics, this characteristic of the shape of the distribution of the mean is especially important.

> If the distribution [of the dependent variable] resembles closely a normal distribution (perhaps, symmetric, unimodal, and slightly platykurtic) then moderate sample sizes (say, 5 or 10) will be sufficient to say that the sampling distribution of the mean is "approximately" normal. For other distributions, 30 observations might be required. (p. 100)

Other writers are not so clear about sample size requirements when choosing between parametric and nonparametric tests. For example, in their classic text on nonparametric statistics, Siegel and Castellan (1988) argue that, "If the sample size is very small, there may be no alternative to using a nonparametric statistical test unless the nature of the population distribution is known exactly" (p. 35). Unfortunately, the authors do not state specifically what they mean by a "very small" sample size. Other authors have discussed implications of using parametric statistical tests when sample sizes are *exceedingly* or *extremely* small (Hays, 1994; Neter et al., 1993), but again, these limits are not specifically defined. Where does that leave us, as researchers, when we are faced with a total sample size of 20 and no likelihood of increasing the size? Do we use parametric or nonparametric tests to analyze our data?

The answer to this question is that "It depends." That is, the choice of parametric or nonparametric test depends on the variables' level of measurement, the shape of the distributions of the variables of interest, and the researcher's knowledge of the variables' distribution in the population. In addition, the researcher needs to consider the number of subgroups to be examined in the analyses. For example, dividing our sample of 20 children into four subgroups depending on the type of minor surgery the children were undergoing (e.g., tonsillectomy, appendectomy, ear surgery, and hernia repair) will result in considerably smaller sample sizes ($n = 5$) to analyze than will merely breaking the group down by experimental and control groups ($n = 10$). Moreover, parametric analyses of the interaction of these two variables (group by type of minor surgery) would be rendered impossible given the resulting small and unequal sample sizes ($n = 2$ or 3 per group) within each subcategory of group by type of surgery interaction.

Equal Numbers of Subjects Within Subgroups

As a general rule of thumb, it is highly desirable to have equal numbers of subjects within subgroups. Although this goal is attainable in experimental and quasi-experimental designs, it is often violated in descriptive studies because of natural selection. It is, therefore, a topic that cannot be avoided in health care research.

Unequal cell sizes have a differential impact on the results of a research study depending on the statistics selected. For example, the problems raised by unequal numbers of subjects are not disastrous when using t tests or simple one-way ANOVAs, particularly if the condition of homogeneity of variance is met (Tabachnick & Fidell, 1996). Serious problems regarding unequal cell sizes with or without homogeneity of variance arise, however, in factorial designs, such as two-way ANOVA, ANCOVA, MANOVA, repeated-measures ANOVA, and multiple regression. These problems are generated because the hypotheses that test the main effects and interaction terms are no longer orthogonal (i.e., independent) when there are unequal cell sizes. Tabachnick and Fidell (1996) offer an in-depth discussion of both the problem of unequal cell sizes and possible solutions in their text on multivariate statistics. The problem of confounding of main effects for multiple factors when there are unequal cell sizes is present in some non-parametric analyses as well.

Reporting Testing Assumptions and Violations in a Research Report

The reporting of assessment of assumptions and their violations does not appear to be a common practice in health care research. Pett and Sehy (1996) found that less than 20% of the 238 randomly selected nursing research articles that they reviewed included even a brief discussion (i.e., a sentence or two) of the assumptions underlying the use of the statistical tests reported. Discussion of the formal testing of the assumptions under-lying the statistical tests, management of violations of these assumptions, and handling of extreme data was also uncommon (15.5%, 14.7%, and 3.4%, respectively). Similar findings were reported by Gaither and Glor-feld (1983) in organizational behavior. These authors offer the hope that researchers do indeed consider the assumptions underlying their choice of statistical tests carefully but either choose not to report their findings or are discouraged to do so by reviewers or editors of journals.

The reporting of tests of assumptions and handling of violations does not have to be an arduous undertaking. A mere sentence or two would suffice. Robichaud-Ekstrand (1991), for example, does this very nicely in her report of responses to shower versus sink baths for 30 patients with myocardial infarction:

> Because the distribution of the overall subjective scores was skewed, a Friedman test ($p \leq .05$) and Wilcoxon post hoc pairwise tests ($p \leq .02$) were used. Pearson [product moment] correlations were used to compare the skewed scores of RPE and the overall subjective scores with the normally distributed HR and BP scores. (p. 378)

Summary

This chapter has attempted to provide the reader with some guidelines for determining which statistics, parametric or nonparametric, are best

---

Box 3.1 Checklist for Assessing the Characteristics of Data

---

1. Determine the levels of measurement of the variables of interest.
2. Evaluate the distribution of these variables:
   - Compare the measures of central tendency for each variable
   - Determine skewness and kurtosis
   - Visually assess the distributions
   - Examine the probability plots
   - Transform the variables if necessary
   - Examine and interpret the results of the transformation
3. Check for homogeneity of variance.
4. Assess the total sample size and size of the subgroups.
5. Determine which statistic, parametric or nonparametric, is best suited for these data.
6. Report the process of evaluation and decisions in the data analysis section.

---

suited for the data at hand. Box 3.1 presents a summary checklist of the procedures that the researcher can use to assess the characteristics of the data being analyzed. These steps include determining the level of measurement of the variables of interest, evaluating their distributions, assessing homogeneity of variances, considering sample sizes, determining the statistics that are best suited for these data, and duly reporting the results of this investigation. It is rare that data collected from the "real world" are perfectly suited to the requirements of a particular parametric or nonparametric test. Remember, however, that the bottom line to the choice of most parametric statistics is the important question: *Do the mean and standard deviation truly represent your data?*

4 "Goodness-of-Fit" Tests

- **Binomial test**
- **Chi-square goodness-of-fit test**
- **Kolmogovov-Smirnov one-and two-sample tests**

In this chapter, we will examine nonparametric "goodness-of-fit" tests. These tests are used when the researcher has obtained a sample of values and wants to know if this sample comes from a specified population distribution. For example, we might want to know whether the proportion of men and women obtained in a collected sample is similar to that in a target population. We might also like to test statistically whether a variable's distribution is similar to that of a normal or other distribution. These types of tests are described as *goodness-of-fit* tests because they compare the results obtained from a given sample to a prespecified distribution. The following tests that will be examined in this chapter are the binomial test, the chi-square goodness-of-fit test, and the Kolmogorov-Smirnov one- and two-sample tests for normal and other distributions.

The Binomial Test

Many nominal-level variables are dichotomous. The variable "gender," for example, consists of two levels: male and female. In our hypothetical intervention study from Chapter 1, we created a grouping variable that had two classes: intervention and control. A third type of dichotomous variable could be outcome following an invasive intervention: cured or not cured.

Suppose we are interested in examining whether the proportion of 20 families in our hypothetical study who are ethnic minorities ($n = 5$) is similar to that routinely seen in the hospital setting from which our sample has been drawn. The binomial test can provide us with this information.

59

This test uses the binomial distribution to determine the likelihood of obtaining a number as extreme as, or more extreme than, that which we obtained in our sample (e.g., $n \leq 5$), given what we know about the numbers of ethnic minority families who routinely use the hospital facilities.

An Appropriate Research Question for the Binomial Test

The binomial test has been used in a number of different ways in the heath care research literature. For example, Cammu and Van Nylen (1995) used this statistic to determine outcomes of pelvic floor muscle exercises after 5 years in 48 women with troublesome incontinence. Harsham, Keller, and Disbrow (1994) also used the binomial test; they compared the growth patterns of 31 infants exposed to cocaine and other drugs in utero with the expected growth of infants in three reference populations. In our hypothetical intervention study, we are interested in comparing the ethnicity characteristics of our sample of convenience with those of the population of families who typically use the study hospital. A research question that could be answered using the binomial test is as follows:

Is the proportion of ethnic minority families who appear in the intervention study similar to the proportion of ethnic minority families who typically use the hospital facilities?

Null and Alternative Hypotheses

Table 4.1 presents the null and alternative hypotheses associated with this research question that are appropriate for the binomial test. Notice that these hypotheses focus on the proportions that would be expected under the null hypothesis that there is no difference between the proportions of ethnic minority families appearing in the sample and in the population. In statistical hypothesis testing, the alternative or research hypothesis can be directional or nondirectional. In our hypothetical intervention study, because our research question is nondirectional, the alternative hypothesis will also be nondirectional (H_a: $p \neq p_0$); that is, the sample will not conform if either too many or too few of the families are in the "ethnic minority" group. If prior research or other sources had offered sufficient evidence, a directional hypothesis could be stated (e.g., H_a: $p < p_0$).

Table 4.1 Example of Null and Alternative Hypotheses Appropriate for a Binomial Test

---

Null Hypothesis

H_0: There is no difference between the proportion of ethnic minority families in our sample compared with that of the hospital in general; that is, $p = p_0$.

Alternative Hypothesis

H_a: There *is* a difference between the proportion of ethnic minority families in our sample compared with that of the hospital in general; that is , $p \neq p_0$.

Overview of the Procedure

To use the binomial test, the data under examination must consist of a single nominal-level variable that is dichotomous. If the variable of interest is not dichotomous (e.g., an "ethnic status" variable that contains more than two classes), it needs to be collapsed carefully into two mutually exclusive categories (e.g., minority and nonminority). Defining the criteria used to define these two distinct classes, however, is the task for the researcher, not the computer. Values of "1" are assigned to those individuals who make up the category of interest (e.g., 1 = ethnic minority), and values of "0" are assigned to those individuals who belong to all other categories (e.g., 0 = ethnic nonminority).

Next, the proportion (p) of subjects who reside in the population of interest who would have been assigned a value of 1 for the target variable had they taken part in the study needs to be determined. Because it is a proportion, this value can take on values only between 0 and 1. Finally, the proportion of subjects in the target population $(1 - p = q)$ who would have been assigned a value of 0 for the target variable is identified. Values for p and q could be obtained from a variety of sources, such as official records, census data, and prior research. In our hypothetical example, we might have patient census information available for the previous 12 months that indicate that the proportion of ethnic minority families who have used the hospital facilities is one third. Our p, therefore, would equal .33 and $1 - p = q = .67$.

The binomial test is based on the binomial distribution. This distribution enables the researcher to determine the probability that the given sample of subjects (e.g., 20 families of whom 5 are ethnic minorities and 15 are nonminorities) could have come from a binomial population whose values of p and q are similar to those of the target population (e.g., .33 and .67). The formula for a binomial distribution is:

$$P[Y = k] = \binom{N}{k} p^k q^{N-k}$$

where

$P[Y = k]$ = The probability of obtaining exactly k observations in one class and $N - k$ observations in the other

k = the number of subjects in the sample who were assigned the value of 1 on the target variable.

$$\binom{N}{k} = N \text{ choose } k = \frac{N!}{(k!) \, (N-k)!}$$

$N =$ the total sample size.

The exclamation point (!) indicates a factorial; for example, $N! = (N)$ $(N - 1)(N - 2) \ldots (2)(1)$. Y represents a random variable that can take on any number of values that are greater than 0 but no larger than the sample size, N.

Suppose that we want to know the probability of obtaining exactly 5 minority families in our study given $p = .33$. This random variable, Y, would take on the value of 5, and the probability that a random sample of size $N = 20$ will contain exactly 5 minority families would be calculated as follows:

$$P[Y = k] = \binom{N}{k} p^k q^{N-k} = P[Y = 5] = \binom{20}{5} .33^5 \, .67^{15} =$$

$$\frac{20!}{5!15!} (.0039) (.0025) = .1493.$$

The probability, therefore, that we would obtain exactly 5 ethnic minority families in a sample of 20 is .1493.

In hypothesis testing, we are not usually interested in the probability of obtaining *exact* values; instead, we often want to know the probability of obtaining values *as extreme as or more extreme than* our obtained value. In our intervention study, for example, we might want to know the probability of obtaining 5 or fewer ethnic minority families in a sample of size 20 given the proportion of such families in the target population (i.e., p = .33). In this case, we would want to calculate P[$Y \le k$], which would be the sum of the probabilities that $Y = 0, 1, 2, 3, 4$, and 5 , or P[$Y = 0$] + P[$Y = 1$] + P[$Y = 2$] + P[$Y = 3$] + P[$Y = 4$] + P[$Y = 5$].

Performing such calculations by hand could entail a lot of pencil pushing. Fortunately, for those without computers, tables of probabilities for the binomial distribution are available for small sample sizes ($N \le 25$) in most statistics textbooks (e.g., Daniel, 1990). For example, if we were to use a table for the binomial distribution for $N = 20$ and p = .33 (e.g., Table A.1, Daniel, 1990), we would find that the one-tailed p value, P[$Y \le 5$] = .0003 + .0033 + .0153 + .0453 + .0947 + .1493 = .3082, is exactly equal to the one-tailed significance level obtained in the computer printout in Table 4.2 (⑤). If this one-tailed p value is less than a one-tailed alpha, the null hypothesis that the sample proportions are equal to those of the population will be rejected. In our hypothetical study, we have a two-tailed α = .05; therefore, we need to multiply our obtained one-tailed p value by 2 and compare it to .05. Because this two-tailed p value (.3082 × 2 = .6164) is considerably greater than α = .05, we will fail to reject the null hypothesis. (Note that because the binomial distribution is not symmetric for p ≠ .5, this procedure for obtaining two-tailed p values may not be completely accurate for binomial distributions with extreme proportions, that is, those close to 0 or 1.)

Using the *z approximation to the binomial distribution.* The binomial distribution also approximates a normal z distribution as N becomes large and p is not too close to the extreme values of 0 or 1. This z distribution can be used when Np and Nq (where q = 1 − p) are both greater than 5 (Daniel, 1990). If this approximation is used in place of the binomial distribution, $P(Y \le k)$ is obtained by examining the probability of obtaining a z statistic as extreme as or more extreme than the following:

$$z = \frac{(Y \pm .5) - Np}{\sqrt{Npq}}$$

($Y + .5$) is used when $Y < N$p and ($Y − .5$) is used when $Y > N$p.

In our hypothetical study, we could use this approximation because we meet Daniel's (1990) criteria (Np = (20)(.33) = 6.7 and Nq = (20)(.67) = 13.3). We would also use the value ($Y + .5$) to calculate the z statistic because Np = 6.7 is greater than 5. Our z statistic, therefore, would be calculated as follows:

$$z = \frac{(Y + .5) - Np}{\sqrt{Npq}} = \frac{(5+.5) - 20(.33)}{\sqrt{(20)(.33)(.67)}} = -.52.$$

Using a table of values for the standard normal distribution that is available in most textbooks on statistics (e.g., Table C.1, in Neter et al., 1993) , we can see that the area under the curve that lies to the left of $z = -.52$ is .3015. Except for approximation error, this one-tailed p value is similar to that which we obtained from the exact values of the binomial distribution (.3082). Again, the two-tailed value, $p = 2(.3015) = .6030$, is considerably larger than our $\alpha = .05$. We must, therefore, fail to reject the null hypothesis and conclude that our sample is not significantly different from the target population. The meaning of these results will be discussed in greater detail when we examine the computer printout.

Critical Assumptions of the Binomial Test

There are not many assumptions underlying the binomial test. This makes it a valuable test to use when other, more powerful tests are not feasible. The three basic assumptions of the binomial test are described below.

1. The randomly selected observations are independent and obtained from a single sample. This assumption means that there are no duplicate observations in which a respondent's answers are counted twice and no respondent has exerted an undue influence on the other responses. In addition, there is a single sample that consists of all the respondents who are taking part in the study. The binomial test is not appropriate for repeated observations obtained from the same sample or when two or more groups are being compared.

2. The data must be in two discrete categories to which the values of 0 and 1 have been assigned. It may be that the variable of interest is at the nominal level of measurement but not dichotomous. If that is the case, the multiple levels of the variable need to be collapsed into two mutually exclusive categories. Better yet, a new variable could be created from the old one, thus retaining the old variable's values for future reference. In our hypothetical example, we may have had several types of ethnic minority families, requiring collapsing of the data into a single category for minorities. In SPSS for Windows, a new dichotomous variable containing two

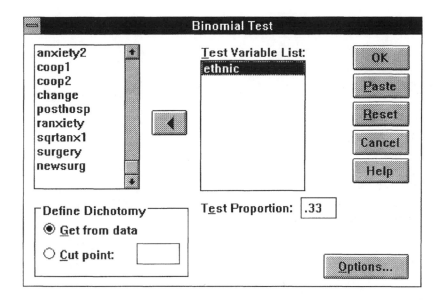

Figure 4.1. Computer Commands for the Binomial Test in SPSS for Windows

levels can be created easily from a multilevel variable by highlighting the commands *Transform . . . Recode . . . Into a Different Variable* in the menu.

3. The probability of an event occurring in a given population can be specified (e.g., p = .5). To use the binomial test, the value of p, or the proportion of events in the population that is likely to take on the value of 1 for the variable of interest, must be specified or estimated. As indicated, these theoretical proportions can come from a variety of sources, such as public records, census data, or prior research.

Computer Commands

The computer commands for the binomial test in SPSS for Windows are generated easily by highlighting the *Statistics . . . Nonparametric Tests . . . Binomial* commands in the menu (Figure 4.1). The theoretical proportion of subjects in the population who would have received the value of 1 for

Table 4.2 Computer-Generated Printout in SPSS for Windows for the Binomial Test

```
RECODE
  ethnicty
  (1=0)  (2 thru Highest=1)  INTO  ethnic . ①
VARIABLE LABELS ethnic 'ethnic minority?'.
VALUE LABELS ethnic 0 'No' 1 'Yes'.
NPAR TEST                              ②
  /BINOMIAL (.33)= ethnic
  /MISSING ANALYSIS.

- - - - - Binomial Test

    ETHNIC    ethnic minority?


    Cases
                              Test Prop. =   .3300 ④
③     5    = 1.00           Obs. Prop. =   .2500
     15    = .00
     --                     Exact Binomial
     20    Total            1-Tailed P =   .3082 ⑤
```

the target value under the null hypothesis is specified with the *Test Proportion* subcommand in the binomial test dialogue box. The default test proportion value for SPSS is 0.5. If this default value is used, a two-tailed probability is presented in the output; otherwise, a one-tailed probability is displayed (Norusis, 1995a).

Computer-Generated Output

Table 4.2 presents the syntax commands and computer-generated printout obtained from SPSS for Windows. The syntax commands indicate that a dichotomous variable, *ethnic*, has been created from the multilevel categorical variable *ethnicty* (①). Next, the syntax commands for the binomial test are presented (②). These commands could be used in lieu of using the menu or may be used with SPSS V5.0 if the Windows version is not available.

Unless otherwise indicated, handling of missing values for nonparametric statistics in SPSS for Windows is specified within the *Options* subcommand in the nonparametric dialogue box. There are two available options for dealing with missing values in this menu: Missing values are excluded from the analysis on a test-by-test basis (the default option), or cases with missing values on *any* variable specified on the *Test Variable* list are excluded automatically from *all* specified analyses. Care should be taken when selecting the second option because this choice could result in a dramatic reduction in the number of observations available for analysis.

The results of the binomial test presented in Table 4.2 show that there were 5 ethnic minority families and 15 nonminority families among the 20 families who participated in this study (③). The printout also indicates that whereas we predicted that the expected proportion of families who would be ethnic minority was .33. (④), the actual proportion of ethnic minority families that we obtained was .25 (5/20 = .25). Is this proportion sufficiently different from our expected proportion to warrant rejecting the null hypothesis?

To determine whether the difference between the proportions, .33 and .25, is large enough to reject the null hypothesis of no difference in proportions, we need to turn to the *p* value in Table 4.2 (⑤) and compare it to our stated alpha level. If the obtained *p* value is less than our alpha, we will reject the null hypothesis of no difference in proportions. If it is greater than our stated alpha, we will fail to reject the null hypothesis.

The obtained *p* value of .3082 presented in the computer output (⑤) is a one-tailed *p* value because our critical test proportion was equal to .33, not .50. It is also similar to what we obtained when using a table of probabilities for the binomial distribution. A one-tailed *p* value is used when the alternative hypothesis is directional. If our alternative hypothesis had been one-tailed (e.g., $p < p_0$), we would compare $\alpha = .05$ to our obtained one-tailed significance level. The decision rule is that we would reject the null hypothesis of similarity of proportions if the obtained one-tailed significance level is less than .05.

Our alternative hypothesis is two-tailed because we are not predicting the direction of difference in proportions. To obtain a two-tailed comparison, we need to double the obtained *p* value (.3082 × 2 = .6164) and compare it to our prestated alpha ($\alpha = .05$). Because the obtained two-tailed *p* value of .6164 is considerably larger than .05, we will retain the null hypothesis. Our conclusion is, therefore, that there is no significant difference between the proportion of ethnic minority families in our sample and that in the hospital in general.

Should we be pleased with or concerned about this decision not to reject the null hypothesis? In this particular situation, we might be pleased to note

Table 4.3 Comparison of the Numbers and Proportions of Ethnic Minority Families in the Study and in the Hospital Population

	Observed		Expected		
Ethnic minority?	N	Proportion	N	Proportion	two-tailed p
Yes	5	.25	6.3	.33	.62
No	15	.75	13.7	.67	
Total	20	1.00	20.0	1.00	

that our sample is not unlike the hospital population, at least with regard to the proportion of ethnic minority families participating in our study. This finding increases the potential generalizability of our findings.

Presentation of Results

There are several approaches that could be taken to present the data obtained from a binomial test. A very simple but effective approach would be to present the results in text form as follows:

> No statistically significant differences were found between the proportion of ethnic minority families who participated in this study and that which would have been expected (p = .33) given the numbers of minority families who utilize the hospital's services.

A second approach might be to present these results in tabular form similar to Table 4.3.

Advantages and Limitations of the Binomial Test

The binomial test is extremely useful in many different types of situations because all that is required is a dichotomous variable and knowledge about the expected proportions in the population. Data also could be collapsed to obtain these dichotomous categories. Conover (1980) indicates that with a bit of ingenuity, the binomial test can be adapted to nearly any data set. He also argues that it is "simple to perform, simple to explain, and sometimes powerful enough to reject the null hypothesis when it should be rejected" (p. 96). Siegel and Castellan (1988) caution, however, that when a continuous variable is dichotomized (e.g., collapsing a person's age into two categories such as under 30 and 30 and over), the choice of the binomial test may result in a waste of rich information contained in the data.

Alternatives to the Binomial Test

Because the binomial test uses only categorical data, there is no parametric counterpart to this nonparametric test. A nonparametric alternative is the chi-square goodness-of-fit test.

Examples From Published Research

Cammu, H., & Van Nylen, M. (1995). Pelvic floor muscle exercises: Five years later. *Urology, 45*, 113-117.

Harsham, J., Keller, J. H., & Disbrow, D. (1994). Growth pattern of infants exposed to cocaine and other drugs in utero. *Journal of the American Diet Association, 94*, 999-1007.

The Chi-Square Goodness-of-Fit Test

Nominal-level variables often are not dichotomous but rather have multiple levels. For example, we may want to be more specific regarding a family's ethnic origins rather than simply distinguishing between ethnic minority and nonminority families. We may also be interested in comparing the types of minor elective surgery scheduled for the children in our sample with what we would have expected given what we know about the hospital population of children in general.

The chi-square goodness-of-fit test (Cochran, 1952) is a useful tool to analyze such data. This chi-square test compares the frequencies for a categorical variable actually obtained from a sample with what would have been expected given what we know or hypothesize about the target population.

An Appropriate Research Question for the Chi-Square Goodness-of-Fit Test

The chi-square goodness-of-fit test is particularly appropriate when researchers want to compare their sample of categorical values to what is known or hypothesized about a target population. It also has been used in

Table 4.4 Example of Null and Alternative Hypotheses for the Chi-Square Goodness-of-Fit Test

---

Null Hypothesis

H_0: There is no difference between the obtained frequencies of minor surgery scheduled for the children in our sample and the expected frequencies obtained from the hospital in general.

Alternative Hypothesis

H_a: There *is* a difference between the obtained frequencies of minor surgery scheduled for our sample of children and the expected frequencies obtained from the hospital in general.

---

in combination with logistic regression to evaluate rates of exposure to abuse, disease, or mortality in various populations (Ratner, 1995; Ruttimann & Pollack, 1991; Zahniser, Gupta, Kendrick, Lee, & Spirtas, 1994). Kelly-Hayes and colleagues (1995) also used the chi-square goodness-of-fit test to examine temporal patterns of stroke onset among the 5,070 people who took part in the Framingham study. In an interesting study that underscores the need for greater use of nonparametric tests in health care research, Gaddis and Gaddis (1994) used the chi-square goodness-of-fit test to evaluate the shape of the distributions of the Glasgow Coma Score (GCS) and Revised Trauma Score (RTS) in a Level I trauma center. The authors concluded that the GCS and RTS distributions were not normally distributed for all data sets that they examined.

In our hypothetical intervention study, we were interested in comparing the types of minor surgery experienced by our sample of convenience with those of the hospital population in general. A research question that could be answered using the chi-square goodness-of-fit test is as follows:

To what extent are the types of minor elective surgery scheduled for the 20 children in the sample representative of the target population of children who are hospitalized for minor surgery?

Null and Alternative Hypotheses

Table 4.4 presents the null and alternative hypotheses for this research question that can be used with the chi-square goodness-of-fit test. The null hypothesis will be rejected if the probability associated with the computed value of the chi-square statistic with $df = k - 1$ is less than the stated level of alpha (e.g., $\alpha = .05$). Note that the hypotheses reflect *frequencies* of observed and expected data and that the alternative hypothesis is nondirectional, or two-tailed.

Overview of the Procedure

The chi-square goodness-of-fit test requires that the data being analyzed be at a nominal level of measurement, with two or more levels. These data must also be presented in the form of frequencies, not scores. The researcher also needs to have available information concerning what frequencies would have been expected given what is known about a particular population.

A chi-square statistic is used to examine differences between observed and expected frequencies generated for each category or cell. That is,

$$\chi^2 = \sum_1^k \frac{(O_i - E_i)^2}{E_i}$$

where

O_i	= observed frequencies in the ith cell
E_i	= expected frequencies in the ith cell
k	= the number of levels of the categorical variable
df	= $k - 1$.

If the sum of the differences between the observed and expected values across all the cells of the variable is sufficiently large, the resulting chi-square value with $k - 1$ degrees of freedom will also be large, increasing the likelihood that the null hypothesis will be rejected.

Critical Assumptions of the Chi-Square Goodness-of-Fit Test

1. The variable of interest must be categorical with two or more mutually exclusive and exhaustive levels. Like the binomial test, the chi-square goodness-of-fit test requires nominal-level data for a single variable that are reflected in frequencies of data, not scores. Unlike the binomial test,

this test can be used with data that have more than two levels of mutually exclusive categories.

In our hypothetical example, the categorical variable of types of minor elective surgery scheduled for the children has four levels that have the following corresponding values: 1 = tonsillectomy, 2 = appendectomy, 3 = ear surgery, and 4 = hernia repair. Each child was assigned to one and only one of the four categories, and the data are expressed in frequencies, not scores on a particular test.

2. The randomly selected observations are independent. Like the binomial test, the chi-square goodness-of-fit test assumes that the randomly selected observations are independent; that is, there cannot be duplicate observations from the same respondent, and one set of observations should not unduly influence another set.

3. The expected frequencies for the variable in the target population must be specified. As for the binomial test, it is necessary to specify the expected frequencies for the chi-square goodness-of-fit test. Expected values for a particular variable are obtained from the knowledge the researcher has regarding the population of interest. In our hypothetical example, we would compare the obtained frequencies of minor surgery for our sample of children with those of the entire hospital population of children who were scheduled for minor surgery during a particular time period to determine whether our sample could possibly have been obtained from such a population.

Suppose that we knew from past records that the expected percentages for the types of minor surgery scheduled for the children were as follows: tonsillectomies = 50%, appendectomies = 20%, ear surgery = 25%, and hernia repair = 5%. These values (50, 20, 25, and 5) would represent our expected values. Note that these values add up to 100. They can also sum to 1.00 if presented in proportional form (e.g., .50, .20, .25, and .05).

4. If the categorical variable is dichotomous, the expected frequencies for each cell should be at least 5. If there are more than two levels for the categorical variable, no more than 20% of the cells should have expected frequencies of less than 5, and no cell should have an expected frequency of less than 1. If the variable of interest is dichotomous, the degrees of freedom associated with it is 1 ($df = k - 1 = 2 - 1 = 1$). For dichotomous variables, the expected value for each of the two cells must be at least 5. This requirement suggests that a minimum sample size for a dichotomous variable needs to be 10 if the expected proportion is equal to .50 for each cell and greater than 10 if the proportions are unequal.

If the nominal-level variable of interest has more than two levels, such as our variable "types of minor surgery," no more than 20% of the cells should have expected frequencies of less than 5, and no cell should have an expected frequency of less than 1. This rule of thumb suggests that a minimum sample size for a categorical variable with four levels when equal proportions are expected is 20; the minimum sample size would increase as the proportions become more disparate.

The reason for this requirement is that the chi-square goodness-of-fit test is only asymptotically chi-square; that is, it approaches a chi-square distribution as the expected frequencies become larger. Monte Carlo simulation studies have indicated that the approximation is sufficient when the expected values are greater than 5 (Gibbons, 1985). Hays (1994) suggests that the expected frequency for each cell should be 10 or more when the degrees of freedom are equal to 1 (i.e., the variable is dichotomous) and should be equal to 5 or more if the degrees of freedom are greater than 1 (i.e., the variable has more than two levels). Note that it is the *expected values*, not the actual values, that need to meet this requirement.

Should the results of the statistical analysis indicate that the data fail to meet this requirement, it is possible to collapse the data to create a smaller number of cells. Such collapsing of cells, however, needs to be undertaken carefully so that the combined cells make intuitive sense. For example, if the categories were ordered (e.g., level of income), it would be preferable to collapse adjacent cells rather than those at the extremes. Care should also be taken when collapsing data, because the procedure may reduce information available and obscure potentially important findings.

Computer Commands

The chi-square goodness-of-fit test can be generated in SPSS for Windows by highlighting the *Statistics . . . Nonparametric Tests . . . Chi-Square* commands in the menu (Figure 4.2). The default option in SPSS for Windows is that the expected frequencies are equal. Unequal expected frequencies can be specified through the *Expected Values* command (Figure 4.2) and can be listed as either proportions, percentages, or actual values (e.g., 50, 20, 25, and 5) in ascending order of the values of the categorical variable (e.g., 1, 2, 3, and 4). Actual values will be reinterpreted as expected proportions. These proportions (e.g., .50) are then multiplied by the total

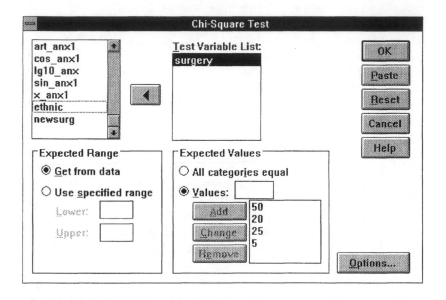

Figure 4.2. Computer Commands for the Chi-Square Goodness-of-Fit Test

number of cases, N, to arrive at the expected number of cases in each corresponding category (Norusis, 1995).

Computer-Generated Output

Table 4.5 presents the SPSS for Windows syntax and computer-generated printout for the chi-square goodness-of-fit test. These syntax commands (①), which also can be used with earlier versions of SPSS, indicate that the chi-square goodness-of-fit test will be undertaken using the variable "surgery" and that the expected percentages for the four categories of this variable in ascending order are 50, 20, 25, and 5. The subcommand "/missing analysis" indicates that the default option for missing values has been selected: Missing values will be excluded on a test-by-test basis. For our hypothetical study, it does not matter which missing values option we

Table 4.5 Computer-Generated Printout for the Chi-Square Goodness-of-Fit Test for Uncollapsed Cells

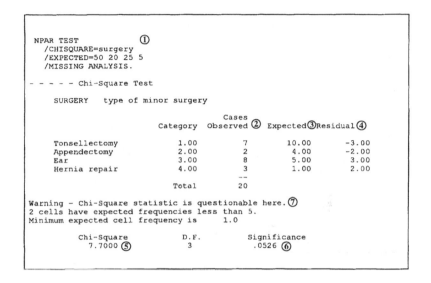

```
NPAR TEST          ①
  /CHISQUARE=surgery
  /EXPECTED=50 20 25 5
  /MISSING ANALYSIS.

- - - - - Chi-Square Test

   SURGERY    type of minor surgery

                                   Cases
                       Category  Observed ②  Expected③Residual④

    Tonsellectomy        1.00         7        10.00     -3.00
    Appendectomy         2.00         2         4.00     -2.00
    Ear                  3.00         8         5.00      3.00
    Hernia repair        4.00         3         1.00      2.00
                                     --
                        Total        20

Warning - Chi-Square statistic is questionable here. ⑦
2 cells have expected frequencies less than 5.
Minimum expected cell frequency is        1.0

         Chi-Square           D.F.        Significance
          7.7000 ⑤             3             .0526 ⑥
```

select because there is only one variable listed in the *Test Variable* list (surgery), and we have no missing values in this study. The output identifies the observed numbers of children in the sample who had been scheduled for tonsillectomies ($n = 7$), appendectomies ($n = 2$), ear surgery ($n = 8$), and hernia repair ($n = 3$), Table 4.5, (②), along with the numbers of children who would have been expected to be scheduled for such minor surgery given our hypothesized proportions (③). These numbers are generated from the proportions presented in the syntax commands for this test (e.g., [.50][20] = 10). The residual that is presented (④) represents the difference between the observed and expected frequencies for this variable (e.g., $7 - 10 = -3$).

As indicated above, the chi-square statistic (χ^2) is generated by summing and averaging the squared differences between the observed (O_i) and expected (E_i) frequencies across all k cells. For our hypothetical surgery data, the χ^2 value was calculated as follows:

$$\chi^2 = \sum_{i=1}^{k} \frac{(O_i - E_i)^2}{E_i} = \frac{(7-10)^2}{10} + \frac{(2-4)^2}{4} + \frac{(8-5)^2}{5} + \frac{(3-1)^2}{1} = 7.70.$$

This generated statistic of $\chi^2 = 7.70$ and its corresponding degrees of freedom, $df = 3$, are presented in the computer printout (⑤).

To determine whether the generated chi-square value of 7.70 is sufficiently large to reject the null hypothesis, the obtained significance level (⑥) needs to be compared to our stated level of alpha (e.g., $\alpha = .05$). If the obtained significance (p) value is greater than the stated level of alpha, the null hypothesis will not be rejected. If the obtained p value is smaller than or equal to the stated level of alpha, the null hypothesis will be rejected. Our obtained p value is .0526, which is just on the *rejection* fence; that is, it is larger than our $\alpha = .05$ but would be equal to .05 if we chose to round off. Technically, we should fail to reject the null hypothesis in this instance.

Note that the computer has warned us that we have violated one of the assumptions of the chi-square goodness-of-fit test because two of the four cells have expected frequencies of less than 5 (⑦). This warning suggests that the results from this particular chi-square test may be questionable and that we need to consider collapsing a few cells. An examination of the computer output suggests that it might be advisable to collapse cells 2 and 4, because they both have expected values of less than 5 (③). Combining these two cells would reduce the number of cells to three and eliminate the expected frequency violation. The question to be answered by the researcher, however, is whether it makes intuitive sense to combine these two cells, appendectomies and hernia repairs, leaving three cells: tonsillectomies, ear surgery, and "other surgeries."

Collapsing Cells to Eliminate Minimum Size Violations

Table 4.6 presents the syntax commands and computer-generated output after having combined cells 2 and 4 using the SPSS for Windows commands *Transform . . . Recode . . . Into the Same Variables*. The syntax commands indicate that levels 2 and 4 of the surgery variable have been recoded (Table 4.6, ①), and the hypothesized population proportions have been altered (50, 25, and 25) to reflect the three levels of the recoded surgery variable (②).

It appears that the collapsing of data was successful, because all the expected values for the type of surgery variable are now 5 or greater (③). The generated chi-square statistic of 2.70 (④), however, is considerably

Table 4.6 Computer-Generated Printout for Chi-Square Goodness-of-Fit
Test After Cells Have Been Collapsed

```
RECODE
  surgery  (1=1) (2,4 = 2) (3=3). ①

EXECUTE .
NPAR TEST
  /CHISQUARE=surgery
  /EXPECTED=50 25 25              ②
  /MISSING ANALYSIS.

- - - - - Chi-Square Test

  SURGERY Type of minor surgery

                                        Cases
                          Category  Observed  Expected  Residual

  Tonsellectomy           1.00          7      10.00      -3.00
  Appendectomy, hernia    2.00          5       5.00 ③     .00
  Ear                     3.00          8       5.00      3.00
                          --
                          Total        20

          Chi-Square          D.F.            Significance
          2.7000  ④             2                .2592  ⑤
```

smaller than the statistic presented in Table 4.5 and is no longer even close
to significance, with a p value of .2592 (⑤). Our decision therefore would
be to fail to reject the null hypothesis. There are no statistically significant
differences between the obtained and expected frequencies with regard to
types of minor elective surgery scheduled for the children.

Presentation of Results

The results of this analysis could be presented in either text or tabular
form. For example, it could be stated in the text:

No statistically significant differences ($p = .26$) were found between the types
of minor surgery scheduled for the children and what would have been
expected given the types of minor surgery ordinarily scheduled for the children
in this hospital.

The data also could be presented in tabular form similar to that in Table 4.7.

Table 4.7 Types of Minor Surgery Scheduled

Type of Minor Surgery	N	Percentage	χ^2	p
Tonsillectomy	7	35.0		
Appendectomy	2	10.0	2.70^1	.26
Ear surgery	8	40.0		
Hernia repair	3	15.0		
Total	20	100.0		

1. df = 2; two cells, Appendectomy and Hernia repair, were collapsed because 2 cells had expected frequencies < 5.

Advantages and Limitations of the Chi-Square Goodness-of-Fit Test

The chi-square goodness-of-fit test can be used to test a variety of problems concerning whether the obtained frequency distribution for a given categorical variable fits a hypothetical theoretical distribution. These theoretical distributions could take on several different forms. Neter and colleagues (1993) outline an approach using the chi-square statistic to test the null hypothesis that the obtained distribution is Poisson. Hays (1994) and Daniel (1990) provide information concerning the approach to testing whether the distribution is normal. In both instances, it is required that the researcher estimate the expected values for such distributions given the number of levels of the discrete (categorical) variable.

In SPSS for Windows, testing of specific distributions is limited to either its default, the uniform distribution (i e., equal frequencies expected in each category), or specified values for the unequal frequencies. A disadvantage of the chi-square goodness-of-fit test is that although, technically, there are numerous theoretical distributions that could be tested using this test, it is necessary to have at hand the actual values of the expected frequencies, and these are not always readily available. Moreover, this test does not arrive at unique results because the value of the statistic depends, to some extent, on the number of levels of the categorical variable being examined (Daniel, 1990; Kallenberg, Oosterhoff, & Schriever, 1985). There are other, more suitable, nonparametric statistics that could be used when a comparison of obtained and theoretical distributions is desired. One such test is the Kolmogorov-Smirnov test for one and two samples. We will examine these statistics later in this chapter.

Alternatives to the
Chi-Square Goodness-of-Fit Test

Given that the chi-square goodness-of-fit test is intended for use with categorical data, there is not a suitable parametric counterpart to this test. Alternative nonparametric tests include the binomial test and the Kolmogorov-Smirnov one-sample test.

Examples From Published Research

Gaddis, G. M., & Gaddis, M. L. (1994). Non-normality of distribution of Glasgow Coma Scores and Revised Trauma Scores. *Annals of Emergency Medicine, 23,* 75-80.
Kelly-Hayes, M., Wolf, P. A., Kase, C. S., Brand, F. N., McGuirk, J. M., & D'Agostino, R. B. (1995). Temporal patterns of stroke onset: The Framingham Study. *Stroke, 26,* 1343-1347.
Ratner, P. A. (1995). Indicators of exposure to wife abuse. *Canadian Journal of Nursing Research, 27,* 31-46.
Ruttimann, U. E., & Pollack, M. M. (1991). Objective assessment of changing mortality risks in pediatric intensive care unit patients. *Critical Care Medicine, 19,* 474-483.
Zahniser, S. C., Gupta, S. C., Kendrick, J. S., Lee, N. C., & Spirtas, R. (1994). Tubal pregnancy and cigarette smoking: Is there an association? *Journal of Women's Health, 3,* 329-336.

The Kolmogorov-Smirnov
One-Sample Test

The chi-square goodness-of-fit test is useful when the researcher has nominal-level data that she or he wants to compare to a theoretical distribution. Sometimes, the variable we wish to compare is a continuous one. The Kolmogorov-Smirnov (K-S) one-sample test is useful for this type of comparison. In SPSS for Windows, it is possible to evaluate whether the distribution of the sample variable is similar to a hypothesized normal, Poisson, or uniform distribution.

At this point, the reader should be familiar with the shape of a normal distribution. Like the normal distribution, the Poisson distribution (Neter et al., 1993) is a family of distributions. Its shapes are determined by a single rate parameter represented by the Greek symbol λ.

The Poisson distribution is often used to compare the likelihood that a number of events (*x*) (e.g., 25 calls to a community crisis hot line) could occur during a fixed time interval (e.g., in a 24-hour period) given a theoretically known Poisson distribution of events (e.g., $\lambda = 0.3$). In our hypothetical study, for example, we might want to know whether the number of postoperative requests for help from the nurse during the first 2 hours following the children's surgery fit a theoretically known Poisson distribution.

The probability function for a Poisson distribution is as follows:

$$P(x) = \frac{\lambda^x e^{-\lambda}}{x!}$$

where

    P(*x*)   = the probability that *x* events will occur during the time interval in question given a theoretical Poisson distribution with rate parameter λ

    λ = the rate parameter of the hypothetical Poisson distribution

    e   = the base of the natural logarithm, which is equal to 2.71828.

λ can assume any positive value between 0 and ∞. Poisson distributions are skewed to the right, but as λ becomes larger, the shape of the distribution becomes more symmetric and eventually approximates a normal distribution.

The uniform distribution, on the other hand, is rectangular and has a uniform or constant probability across each of its values (Neter et al., 1993). This means that the frequencies associated with each value of the variable being examined are equal. The upper (*b*) and lower (*a*) endpoint values of the variable determine the probability function for the uniform distribution:

$$f(x) = 1/(b - a)$$

The uniform distribution has been used in simulation studies and to generate random numbers. It also has been used to evaluate whether the distribution of a phenomenon in a sample is similar across all categories.

In our hypothetical example, we might want to know whether a distribution that looks uniform does, indeed, take on that shape.

An Appropriate Research Question for the Kolmogorov-Smirnov One-Sample Test

The K-S one-sample test is especially useful for assessing the normality of a distribution. As we saw in Chapter 3, this is especially important when we wish to determine whether a parametric statistic is appropriate for a particular analysis. Rosen, Debanne, Thompson, and Dickinson (1992) used this approach to examine distributions in their study of the relationship between abnormal labor and infant brain damage. The K-S one-sample test also has been used to evaluate circadian rhythms in the request for helicopter transport of cardiac patients (Fromm, Levine, & Pepe, 1992), sleep patterns among infants (Haddad, Jeng, Lai, & Mellins, 1987), and the similarity between panic symptom sequence distributions and hypothesized pathophysiologic models (Katerndahl, 1990).

In our hypothetical intervention study, suppose that we wish to determine whether the preintervention anxiety scores of the children in the sample were normally distributed. Our research question for this example would be:

To what extent do the preintervention anxiety scores of the children in our sample resemble a normal distribution?

Null and Alternative Hypotheses

The null hypothesis for the K-S one-sample test is that the cumulative distribution for the sample variable, $F(x)$, is similar to the theoretical cumulative distribution, $F_0(x)$. The alternative, or research, hypothesis may be directional or nondirectional, although it is probably rare that a directional hypothesis would be used. A nondirectional alternative hypothesis merely states that the theoretical and obtained cumulative distributions are disparate, $F(x) \neq F_0(x)$. Table 4.8 presents the null and nondirectional alternative hypotheses that would follow from the research question that we have generated.

Overview of the Procedure

To determine whether the sample distribution could conceivably have come from any given theoretical distribution, the K-S one-sample test

Table 4.8 Example of Null and Alternative Hypotheses Appropriate for the Kolmogorov-Smirnov One-Sample Test

---

Null Hypothesis

H_0: The cumulative distribution of the preintervention anxiety scores for the children in our sample is similar to that of a normal distribution.

Alternative Hypothesis

H_a: The cumulative distribution of the preintervention anxiety scores for the children in our sample is not similar to that of a normal distribution.

---

compares the cumulative distribution of the sample variable with what would have been expected to occur had the sample been obtained from the theoretical parent distribution. In our example, the K-S test compares the cumulative distribution of preintervention anxiety scores of the children in our sample with what would have been expected had the distribution been a perfectly bell-shaped normal distribution. If the point of greatest divergence between the obtained and expected cumulative distributions is sufficiently large, the null hypothesis of normality will be rejected.

The exact methods used to determine the value of the test statistic for the K-S test are more mathematically cumbersome than is desired for presentation in this text. The interested reader is referred to Conover (1980), Daniel (1990), and Siegel and Castellan (1988) for excellent descriptions of the mathematical approach to calculating this statistic.

Critical Assumptions of the Kolmogorov-Smirnov One-Sample Test

1. The sample variable should consist of a randomly selected set of observations that are at least on an ordinal scale of measurement. As with all nonparametric statistics, the K-S one-sample test assumes that the data have been generated from a sample that has been randomly selected. As is often the case in health care research, our hypothetical data were collected from a sample of convenience; therefore, we have violated this assumption. It is important, therefore, to assess the extent to which our nonrandom sample is representative of the general population of hospitalized children. This becomes an issue not only for choice of statistic but also for generalizability.

The K-S one-sample test uses the cumulative distribution function to compare the obtained and theoretical distributions. It is assumed that the variable being examined is continuous. This would imply that the variable being examined has at least an ordinal scale of measurement. Our variable, preintervention anxiety, consists of scores obtained from an ordinal 7-point Likert-type scale, indicating that we have met this assumption.

2. *The theoretical distribution to which the sample variable is being compared must be completely specified.* This assumption requires that for every value of the study variable, the value of the cumulative distribution function can be determined (Siegel & Castellan, 1988). Sometimes, values for a theoretical distribution do not exist (e.g., negative or zero values for a logarithm scale), and therefore the cumulative distribution function for those values cannot be defined. The K-S test could not be used to compare the distribution of a study variable that has negative or zero values to that of a log distribution without first shifting the range of the scale values to exclude the negative and zero values. For example, we could add the value of +4 to every observation in a scale that ranges from –3 to +3 to create an adjusted scale that ranges from +1 to +7.

SPSS for Windows is limited in the number of distributions to which the researcher can compare her or his obtained cumulative distribution: the normal, Poisson, and uniform distributions. The limited availability of comparison distributions may be a result, in part, of the fact that, for distributions that are not determined by the population mean and variance, it is sometimes difficult to estimate, from sample data, the parameters that are critical to defining the distribution.

Computer Commands

There are two ways that the K-S one-sample test can be generated in SPSS for Windows. In Chapter 3, we saw that this test is used in the *Statistics . . . Summarize . . . Explore* dialogue box to assess whether a continuous variable is normally distributed. In addition to the K-S one-sample test, the Shapiro-Wilks test and the Lilliefors significance level for the K-S test are presented in this dialogue box. The Shapiro-Wilks test is similar to the K-S test in that it also tests for normality of distributions. The Lilliefors significance level for the K-S test is used because it provides a better approximation to the normal distribution than does the K-S statistic alone (Conover, 1980). Chapter 3 addressed the interpretation of the computer-generated printout obtained from these commands.

A second way to obtain the K-S one-sample statistic in SPSS for Windows is by highlighting the *Statistics . . . Nonparametric Tests . . . 1-Sample K-S* commands in the menu. These commands enable the researcher to test the hypotheses that the sample distribution is not only normal but also uniform or Poisson. Figure 4.3 presents the *One-Sample K-S* dialogue box in SPSS for Windows. Clicking on the *Test Distribution . . . Normal* box will generate a normal distribution for comparison.

Computer-Generated Output

Table 4.9 presents the syntax commands and computer-generated printout obtained from SPSS for Windows. The syntax commands, which may also be used with earlier versions of SPSS (e.g., Version 5.0), indicate that the normal distribution has been specified (①). Next, the computer-generated printout presents the most extreme absolute, positive, and negative differences between the actual and theoretical (normal) distributions (②). The most extreme absolute difference (.24562) is used to obtain a z statistic (1.0984) (③). If the significance level (p value) of this test statistic is smaller than our stated level of alpha (e.g., $\alpha = .05$), we will reject the null hypothesis that states that the actual and theoretical distributions are similar. Our generated two-tailed p value, .1789 (④), is larger than our stated level of alpha (.05), indicating that we must fail to reject the null hypothesis that our distribution is similar to a normal distribution. Note that this decision is similar to one made in Chapter 3 (Table 3.3, ②). The only difference is that the significance level for Table 4.9 is slightly larger (.1789, ④) than the K-S (Lilliefors) p value given in Table 3.3 (.0572, ②). This discrepancy may be due to the Lilliefors adjustment undertaken in the Chapter 3 analysis.

Presentation of Results

The results of the K-S one-sample test are best presented in text form. For example, they could be stated in the text as follows:

An examination of the preintervention anxiety scores using the Kolmogorov-Smirnov one-sample goodness-of-fit test indicates that the sample data were normally distributed (K-S $z = 1.10$, $p = .18$).

It would be difficult and confusing to present this information in table format.

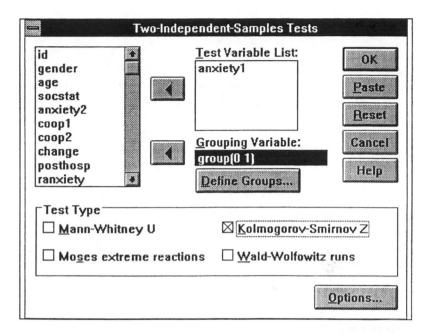

Figure 4.3. SPSS for Windows Dialogue Box for the Kolmogorov-Smirnov One-Sample Test

Table 4.9 Syntax and Computer-Generated Printout for the Kolmogorov-Smirnov One-Sample Test

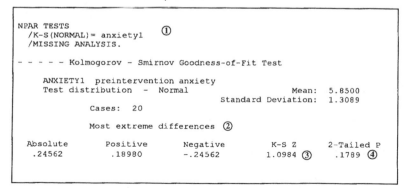

Advantages and Disadvantages of the Kolmogorov-Smirnov One-Sample Test

The K-S one-sample test is especially useful when evaluating the underlying structure of a variable's distribution. It is more powerful than its alternative, the chi-square goodness-of-fit test, especially when sample sizes are small and the level of measurement is ordinal (Goodman, 1954; Lilliefors, 1967; Siegel & Castellan, 1988; Slakter, 1965).

Unfortunately, SPSS for Windows has only three available distributions that can be tested using this statistic: the normal, uniform, and Poisson. Although the normal distribution is the one most commonly compared, the K-S test has been found to be extremely sensitive to even slight deviations from normality, particularly with larger sample sizes. This makes it more likely that the null hypothesis of normality of distributions will be rejected even when, visually, the data distribution appears to be nearly normal. It is recommended, therefore, that the researcher use this statistic only as a beginning assessment of normality. Chapter 3 outlined additional methods for assessing normality.

Alternatives to the Kolmogorov-Smirnov One-Sample Test

There is no parametric counterpart to the K-S test. A nonparametric alternative is the chi-square goodness-of-fit test.

Examples From Published Research

Fromm, R. E., Levine, R. L., & Pepe, P. E. (1992). Circadian variation in the time of request for helicopter transport of cardiac patients. *Annals of Emergency Medicine, 21,* 1196-1199.

Haddad, G. G., Jeng, H. J., Lai, T. L., & Mellins, R. B. (1987). Determination of sleep state in infants using respiratory variability. *Pediatric Research, 21,* 556-562.

Katerndahl, D. A. (1990). Comparison of panic symptom sequences and pathophysiologic models. *Journal of Behavior Therapy and Experimental Psychiatry, 21,* 101-111.

Rosen, M. G., Debanne, S. M., Thompson, K., & Dickinson, J. C. (1992). Abnormal labor and infant brain damage. *Obstetrics and Gynecology, 80*(6), 961-965.

The Kolmogorov-Smirnov
Two-Sample Test

A second type of Kolmogorov-Smirnov (K-S) goodness-of-fit test is the two-sample alternative. This test was originally developed by Smirnov (1939) but subsequently was named the Kolmogorov-Smirnov two-sample test because both Kolmogorov and Smirnov developed similar types of tests (Kolmogorov for the single sample, Smirnov for two samples) (Daniel, 1990).

In this test, the distribution of a continuous variable for two independent samples is compared. The goal of this test is to determine whether the two samples that represent two levels of a categorical variable have been drawn from populations sharing the same distribution (e.g., preintervention anxiety scores for our intervention and control groups). This K-S two-sample test compares the cumulative distributions of the dependent variable within the two categories. If the two cumulative distributions diverge too greatly at any particular point, there may be reason to believe that these distributions do not share the same parent distribution, whatever that distribution might be. Note that, unlike the K-S one-sample test, the two-sample test does not specify the exact form of the two distributions.

Examining the similarity of distributions of a variable within different groups provides the researcher with valuable information. It could be, for example, that although two samples share similar measures of central tendency, the shapes of their distributions are disparate. The K-S two-sample test will examine these discrepancies to determine whether the divergence between the two distributions is sufficiently large to reject the null hypothesis of similarity of distributions. In many ways, the K-S two-sample test is similar to the median, Mann-Whitney, and t tests in that it compares distributions of two independent samples. A huge advantage of the K-S test over these others is its more complete examination of the similarity of cumulative distributions, not merely the measures of central tendency.

An Appropriate Research Question for the
Kolmogorov-Smirnov Two-Sample Test

The K-S two-sample test has been used in numerous ways to compare distributions of two populations. Thornbury (1992) used the K-S two-

Table 4.10 Example of Null and Alternative Hypotheses Appropriate for
the Kolmogorov-Smirnov Two-Sample Test

Null Hypothesis

H_0:   The distributions of the preintervention anxiety scores for the children assigned to the
intervention and control groups in our sample are similar.

Alternative Hypothesis

H_a: The distributions of the preintervention anxiety scores for the children assigned to the
intervention and control groups in our sample are not similar.

---

sample test to compare cognitive performance on Piagetian tasks by 30
Alzheimer's disease patients and their controls. Campbell, Gratton,
Salomone, and Watson (1993) used the same test to evaluate the effects of
physical barriers on response time intervals from ambulance arrival at the
scene to paramedic contact with the patient, whereas Van Lith, Pratt,
Beekhuis, and Mantingh (1992) compared the distributions of immunoreactive inhibins in the maternal serums of 10 Down's syndrome and 80
normal fetus pregnancies.

In our hypothetical intervention study, we might be interested in comparing the similarity of the preintervention anxiety score distributions for
the intervention and control groups. A research question that could be
answered using the K-S two-sample test would be as follows:

To what extent are the distributions of preintervention anxiety scores for the
intervention and control groups similar?

Null and Alternative Hypotheses

Table 4.10 presents the null and alternative hypotheses for the K-S
two-sample test that follow from the research question posed. Note that the
hypotheses do not state the form of the distributions. Unlike the K-S
one-sample test, the two-sample test cannot determine this issue.

Overview of the Procedure

To use the K-S two-sample test, the cumulative frequency distribution
for each sample of observations is determined, using similar intervals for
both distributions. Next, for each interval, the difference between the two

cumulative distribution functions is obtained. The K-S two-sample test then determines the largest discrepancy between the two functions. This discrepancy forms the basis for determining the test statistic.

The K-S two-sample test can be one- or two-tailed, depending on the stated hypotheses; that is, one distribution can be predicted to be different from the other in a particular direction. In our hypothetical example, the alternative hypothesis is nondirectional, indicating that the test is two-tailed.

The mathematical method for determining the significance for this two-sample test depends on the sample size and is offered in Siegel and Castellan (1988) and Conover (1980). SPSS for Windows provides this test among its series of nonparametric statistics.

**Critical Assumptions of the
Kolmogorov-Smirnov Two-Sample Test**

1. The randomly selected dependent variable must be at least ordinal in its level of measurement. Like the one-sample test, the K-S two-sample test derives a statistic that is based on a cumulative distribution. For that reason, the randomly selected dependent variable must be at least ordinal in its level of measurement. The scores for our variable, preintervention anxiety, are derived from a 7-point ordinal scale, suggesting that we have met the assumption of ordinal-level measurement. Our data, however, are from a sample of convenience and are not a random sample.

2. The independent or grouping variable must have two mutually exclusive levels. To determine differences between two groups, it is necessary to have a grouping variable with two mutually exclusive levels. This means that a subject may be assigned to one and only one level of the independent variable. Should this categorical variable have several levels to it (e.g., ethnic status), the multiple levels may be collapsed to two mutually exclusive categories, or the levels may be compared two at a time. Our variable of interest, group membership, has only two levels to it (intervention vs. control group); therefore, we have met this assumption.

Computer Commands

Figure 4.4 presents the dialogue box for the K-S two-sample test that was obtained in SPSS for Windows. This dialogue box was opened by

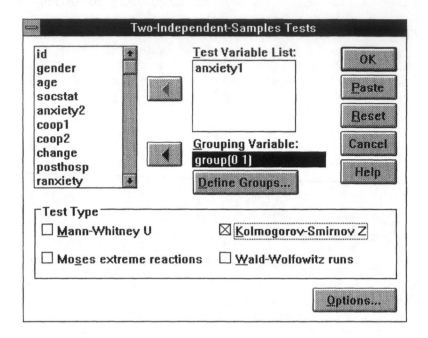

Figure 4.4. SPSS for Windows Dialogue Box for the Kolmogorov-Smirnov Two-Sample Test

highlighting *Statistics . . . Nonparametric Tests . . . Two-Independent-Samples-Tests . . . Kolmogorov-Smirnov z* in the menus. Included in the *Options* command is a request for descriptive statistics, if desired.

As indicated, the K-S statistic cannot determine the shape of the distributions. To determine whether the two sample distributions came from a specified normal distribution, it would be necessary to use the *Statistics . . . Summarize . . . Explore* dialogue box. A grouping variable is then selected (e.g., Group) for evaluations of the subgroups. To obtain assessments of other distributions (e.g., the uniform or Poisson), it would be necessary to first separate out the levels of the independent variable through the *Data . . . Select Cases* dialogue box. It is possible to use conditional expressions to select cases through the *If Condition Is Satisfied* command in SPSS for Windows to isolate the groups (e.g., if Group = 0). A one-sample K-S test for each level of the independent variable could then be undertaken.

Table 4.11 Computer-Generated Printout for the Kolmogorov-Smirnov Two-Sample Test

```
NPAR TESTS
  /K-S= anxiety1    BY group(0 1) ①
  /MISSING ANALYSIS.


- - - - Kolmogorov - Smirnov 2-Sample Test

   ANXIETY1    preintervention anxiety
by GROUP      exper-control grps

       Cases

          10  GROUP =  .00  experimental
          10  GROUP = 1.00  control
          --
          20  Total

    Warning - Due to small sample size,
              probability tables should be consulted. ⑤

         Most extreme differences ②
      Absolute    Positive    Negative        K-S Z ③ 2-Tailed P ④
      .10000      .10000      .00000          .224     1.000
```

Computer-Generated Output

Table 4.11 presents the syntax commands and computer-generated print-out obtained for the K-S two-sample test obtained in SPSS for Windows. The syntax commands (①) can also be used for earlier versions of SPSS (e.g., Version 5.0). As with the K-S one-sample test, we are presented with the most extreme differences between the two distributions, but this time we are comparing the intervention and control groups (②). The most extreme absolute difference between the two distributions, .10000, will be used to generate a K-S z statistic (③). If the two-tailed p value (④) is smaller than our prestated alpha (e.g., $\alpha = .05$), we will reject the null hypothesis that states that the distributions of the intervention and control groups are similar.

Our obtained two-tailed p value, 1.000, is considerably larger than our $\alpha = .05$. In fact, it could not get any larger! Our decision, therefore, is that we must fail to reject the null hypothesis and conclude that the distributions for the preintervention anxiety variable for both groups are similar.

Note that we have been given a warning (⑤) that our sample sizes are small ($n = 10$ in each group) and that we may want to consult probability tables instead of relying on the z statistic, which is based on a normal distribution. Siegel and Castellan (1988) provide tables of critical values (Tables L_I and L_{II}) for assessing the K-S two-sample test when the size of each of the groups is 25 or less. The null hypothesis is rejected if the following inequality holds:

$(n_1)(n_2)$(most extreme absolute difference) > tabled value.

In our hypothetical intervention study, we have 10 subjects in each group, so $n_1 = n_2 = 10$. Our tabled value (Siegel & Castellan, 1988, Table L_I), given a sample size of 10 and two-tailed $\alpha = .05$, is 60. Because the computer printout indicates that the absolute value of the most extreme differences is .1000 (②), our calculated value is (10)(10)(.10) = 10, which is considerably less than the tabled value of 60. The null hypothesis of similarity of distributions of the preintervention anxiety variable for the intervention and control groups therefore cannot be rejected. This result should increase our confidence that, at least at pretest, the intervention and control groups shared similar anxiety.

Presentation of Results

As for the K-S one-sample test, the results of the K-S two-sample test are best presented in text form:

The results of the Kolmogorov-Smirnov two-sample test indicate that the distributions of the preintervention anxiety variable for the intervention and control groups were similar (K-S $z = .22$, $p = 1.00$).

Advantages and Limitations of the Kolmogorov-Smirnov Two-Sample Test

The K-S two-sample test is useful for comparing similarity of two distributions. It has the unique advantage over its parametric counterpart, the t test, of comparing the entire distribution, not just measures of central tendency. This would be a useful test, therefore, if the researcher did not find any differences in the measures of central tendency between the intervention and control groups or from pretest to posttest but suspected that the shapes of the distributions were very different.

A disadvantage of this test is that it is extremely sensitive to unusually large deviations between the two distributions. This increases the likelihood of rejecting the null hypothesis and concluding that there is a difference between the distributions.

Alternatives to the Kolmogorov-Smirnov Two-Sample Test

The parametric counterpart of the K-S two-sample test is the t test for independent means. Alternative nonparametric tests include the chi-square goodness-of-fit test, the median test, and the Mann-Whitney U test. Siegel and Castellan (1988) report that, for small samples, the K-S two-sample test is nearly as powerful as the parametric t test (95% efficiency) but declines somewhat in power as n becomes larger. It is more powerful, however, than either the chi-square test or the median test and is slightly less powerful than the Mann-Whitney U test (see Chapter 6) when samples are large.

Examples From Published Research

Campbell, J. P., Gratton, M. C., Salomone, J. A., & Watson, W. A. (1993). Ambulance arrival to patient contact: The hidden component of prehospital response time intervals. *Annals of Emergency Medicine, 22*, 1254-1257.

Thornbury, J. M. (1992). Cognitive performance on Piagetian tasks by Alzheimer's disease patients. *Research in Nursing and Health, 15*, 11-18.

Van Lith, J. M., Pratt, J. J., Beekhuis, J. R., & Mantingh, A. (1992). Second trimester maternal serum immunoreactive inhibiting as a marker for fetal Down's syndrome. *Prenatal Diagnosis, 12*, 801-806.

Discussion

In this chapter, we have examined four goodness-of-fit tests that are useful for one- and two-sample designs: the binomial test, the chi-square goodness-of-fit test, and the K-S one- and two-sample tests. These goodness-of-fit tests are useful for testing whether the data being examined fit a prespecified distribution or model. The binomial test is used when the categorical variable being examined is dichotomous. It is especially useful for small sample sizes and when the assumptions of the chi-square goodness-of-fit test are not met. The chi-square goodness-of-fit test is used when the sample size is sufficiently large to meet the expected frequency requirements (e.g., expected $n \geq 5$ when $df = 1$). It also can be used when there are multiple levels to the categorical variable.

The chi-square goodness-of-fit test was designed for use with frequency data. When data are continuous (i.e., measurement on at least an ordinal scale), the more powerful K-S one- and two-sample tests can be used. The K-S one-sample test compares the cumulative distribution of the target variable to that of a theoretical distribution (e.g., normal, uniform, or Poisson). The K-S two-sample test, on the other hand, examines the goodness of fit of a cumulative distribution of the dependent variable for two levels of a dichotomous categorical variable. The K-S one- and two-sample tests are considered to be more powerful than either the binomial or the chi-square goodness-of-fit test but do require that the dependent variable be continuous.

5 Tests for Two Related Samples: Pretest-Posttest Measures for a Single Sample

- **McNemar test**
- **Sign test**
- **Wilcoxon signed ranks test**

In Chapter 4, we examined nonparametric tests that could be used to determine the extent to which a single sample is similar to a hypothesized theoretical sample. In health care research, we frequently try to assess whether a particular intervention is effective with a certain population. In our hypothetical intervention example from Chapter 1, we were interested in comparing a group of hospitalized children who received a preoperative intervention designed to promote satisfactory postoperative adjustment with a control group of hospitalized children who received the standard preoperative procedures. In this study, there may be certain characteristics of the sample, such as type of operation to be performed or age and gender of the child, that are known from prior research to confound and potentially misrepresent the outcomes of the intervention.

Two approaches that can be used to address this problem prior to the intervention are the following: (1) matching the subjects with regard to these extraneous confounding variables and then randomly assigning one of the pairs to the control and the other to the intervention group; and (2) using each subject as his or her own control. Data being analyzed then become paired, either through the use of related samples or through repeated observations on a single sample. A third occurrence of matched pairs can occur when the researcher has sampled observations in pairs (e.g., husband and wife). Although each member of the pair may have separate scores on a dependent variable (e.g., marital satisfaction), there is reason to believe that knowing the scores of one member of the pair (e.g., wife's marital satisfaction) will give information about the scores of the other member (husband's marital satisfaction).

There are many types of naturally occurring situations in which respondents are paired, such as mothers and fathers, husbands and wives, twins, and parents and children. A research design could also be longitudinal, with multiple observations of a single sample but lacking a control group. In these situations, a statistical test for independent groups is inappropriate because the data are paired. We are restricted instead to paired tests that acknowledge the dependence of the observations in the samples.

If we were using parametric statistics, the paired t test typically would be used to analyze paired samples. This type of t test is very useful and robust to violations of its assumptions. Sometimes, however, it is not possible to use the paired t test because the data do not sufficiently meet the test's assumptions. Perhaps the sample size is too small, the continuous dependent variable is severely skewed, or the data are categorical. In these situations, the statistical test of choice would be nonparametric.

There are several nonparametric statistical procedures that are suitable for use with paired observations or repeated measures using a single sample across two time periods. Three tests will be examined in this chapter: the McNemar test, the sign test, and the Wilcoxon signed ranks test. Two additional tests for repeated measures across more than two time periods (Cochran's Q and the Friedman test) will be examined in Chapter 6. The McNemar test and Cochran's Q test are used when the measurement of the dependent variable is dichotomous, whereas the remaining tests are used when the dependent variable is at least ordinal in its level of measurement.

The McNemar Test

The McNemar test is especially useful when the researcher has a pretest-posttest design in which the subjects serve as their own controls and the dependent variable is dichotomous (Bennett & Underwood, 1970; Feuer & Kessler, 1989; Siegel & Castellan, 1988). In health care research, for example, we might be interested in comparing the performance of two procedures, comparing the opinions of two experts, or determining whether an educational program altered people's preferences for a particular type of health provider (e.g. nurse practitioner vs. physician). The McNemar test examines the extent of change in the dichotomous variable from pretest to posttest. If the proportion of changed responses in one direction is sufficiently greater than what would be expected by chance, the null hypothesis of no disproportionate change is rejected.

An Appropriate Research Question for the McNemar Test

Several examples from the research literature illustrate the versatility of the McNemar test. Bustamante and Levy (1994) used the McNemar test to evaluate the adequacy of two different procedures to obtain sputum specimens and PCP diagnosis in a group of 28 persons who tested positive for HIV. Using this same test, Snoey and colleagues (1994) compared error rates of physicians and cardiologists regarding their interpretations of emergency department ECGs. Henry (1991) examined the effect of level of patient acuity on clinical decision making of 68 critical care nurses completing two computerized clinical simulations, and Wechsberg-Wendee, Cavanaugh, Dunteman, and Smith (1994) examined the changing needle practices among 287 injecting drug users from community outreach and methadone treatment clinics. In all these examples, the data that were analyzed using the McNemar test were dichotomous variables that were paired either through matched samples or through a design in which the subjects served as their own controls.

In our hypothetical intervention study, the children in the anxiety reduction intervention groups were asked before and after the intervention whether or not they were afraid of their upcoming operations. Because the pretest-posttest measure is dichotomous (the children answered "yes" or "no" to the question concerning fear) and the data are paired, the use of the McNemar test is appropriate. An example of a research question that could be answered using this test would be as follows:

> Is the clinical intervention effective in reducing children's fears of their upcoming operations?

Note that we are examining the pretest and posttest responses of one sample of children who have served as their own controls. We are not comparing two groups.

Null and Alternative Hypotheses

Table 5.1 presents the null and alternative hypothesis for the McNemar test that would follow from the research question. Although nondirectional

Table 5.1   Example of Null and Alternative Hypotheses Appropriate for the
                 McNemar Test

Null Hypothesis

H_0:   Among those children who took part in the intervention and who change their
          reported fearfulness of their upcoming operations, the probability that a child's fear
          is reduced is the same as the probability that the child's fear is increased,

          $P_{reduced} = P_{increased}$.

Alternative Hypothesis

H_a: Among those children who took part in the intervention and who change their
          reported fearfulness of their upcoming operations, the probability that a child's fear-
          fulness is lowered is greater than the probability that the child's fear is increased,

          $P_{reduced} > P_{increased}$.

---

tests are possible, our alternative hypothesis predicts a direction for the probability statement; therefore, the test is directional and we will be using a one-tailed alpha level (e.g., $\alpha = .05$).

Notice that the McNemar test focuses only on those children in the sample who change their opinions regarding their fear and does not include those children whose reported fear does not change. The test also does not compare intervention and control groups, because this is a test of change in a single group.

Overview of the Procedure

To undertake a McNemar test, the data first need to be cast into a 2×2 table that represents the change in an individual's response from before to after the intervention. If the original response data are not nominal, they need to be reduced to a form such that the coding scheme (e.g., 0s and 1s) represents identical values for the paired variables being examined. Table 5.2 presents the form that such a table typically takes.

In our hypothetical example, "A" represents the number of children in the intervention group who reported not being fearful of the operation prior to the intervention but reported being fearful following the intervention (①). "D" represents the number of children in the experimental group whose fears of the operation decreased following the intervention (②). "A + D" therefore is the total number of subjects who changed their responses from before to after the intervention.

Table 5.2 2 × 2 Table for the McNemar Test for Fears of Upcoming Surgery

		After Intervention	
		Yes *(1)*[1]	*No* *(0)*
Expressed Fear			
Before Intervention			
No (0)[1]		A ①	B
Yes (1)		C	D ②

1. The inconsistent ordering of the categories was deliberate, as a means of duplicating the pattern presented on the computer printout.

When sample sizes are reasonably large (i.e., when the total frequency of changes is greater than 10 [Siegel & Castellan, 1988], the McNemar test uses a chi-square test to compare the number of subjects who changed in a specific direction (i.e., A – D) with the frequency of change that would be expected under the null hypothesis of a change being equally likely in both directions ([A + D]/2). This chi-square statistic is similar to the goodness-of-fit statistic that was presented in Chapter 4:

$$\chi^2 = \sum_1^k \frac{(O_i - E_i)^2}{E_i}$$

where

O_i = the observed number of cases in cell i

E_i = the expected number of cases in cell i

k = the number of cells that represent change.

For the McNemar test, we are interested only in the *observed* and *expected* values of two cells, A and D, because these are the only cells that represent change. The *expected* values for each of these cells are (A + D)/2. The chi-square statistic for the McNemar test is given by:

$$\chi^2 = \sum_{1}^{k} \frac{(O_i - E_i)^2}{E_i} = \frac{[A - (A+D)/2]^2}{(A+D)/2} + \frac{[D - (A+D)/2]^2}{(A+D)/2}.$$

If the sum of the differences between the observed and expected values across the change cells is sufficiently large, the resulting chi-square value with 1 degree of freedom ($df = k - 1 = 2 - 1$) will also be large, increasing the likelihood that the null hypothesis will be rejected. When the sample size is smaller (e.g., with expected change frequencies less than 10), a binomial test is used instead (see Chapter 4).

Critical Assumptions of the McNemar Test

There are four major assumptions of the McNemar test.

1. The dichotomous variable being assessed has assigned values for each level (e.g., 0 and 1), the meaning of which is similar across both time periods. This assumption implies that the dichotomous variable for the pretest and posttest is coded similarly. In our hypothetical study, for example, children who expressed no fear concerning the operation were assigned the value of 0, whereas children who did express fear were assigned the value of 1 for both the pretest and the postintervention measure.

2. The data being examined represent frequencies, not scores. As with all chi-square statistics, the assumption is made that the data being examined are frequency data, not scores. In our hypothetical study, the children answered "yes" (1) or "no" (0) to the question of whether they were fearful of their upcoming operations. These data represent frequencies or counts and, therefore, meet this assumption.

3. The dichotomous measures are paired observations of the same randomly selected subjects or matched pairs. It is expected that the data consist of paired responses from a set of randomly selected subjects or matched samples of subjects. In our hypothetical study, the children were not randomly selected because this was a sample of convenience. The observations were paired, however, because the children served as their own controls by responding to the question concerning their fear at two different points in time.

4. The levels of the dichotomous variable are mutually exclusive; that is, a subject can be assigned to only one level of the dichotomous variable that is being examined over time. This assumption implies that a subject cannot be assigned to both a 0 and a 1 on the dichotomous variable at pretest and posttest. In our hypothetical intervention study, a child could not report *both* a "yes" *and* a "no" on the pre- or postintervention fear variables.

Computer Commands

Figure 5.1 presents the SPSS for Windows dialogue box that is used to generate the McNemar test. This dialogue box was opened by choosing the following items from the menu: *Statistics . . . Nonparametric Tests . . . 2 Related Samples.* Because we were interested in restricting the analysis to only those children who received the intervention ($n = 10$), the sample was restricted by first using the *Select Cases* command (*Data . . . Select Cases*) and instructing the computer to select only those cases that met the condition that they had been assigned to the intervention group (i.e., Group = 0).

Next, the pair of dichotomous variables to be examined is selected from the menu. These variables can be either repeated measures from subjects who are being used as their own controls (e.g., fearbef = fear before the intervention and fearpost = fear following the intervention) or dichotomous variables obtained from matched samples (e.g., husbands' and wives' marital satisfaction). It should be noted that, in SPSS for Windows, the selected variable that appears first alphabetically in the variable list will become Variable 1 under *Test Pairs List* despite what is indicated under *Current Selections.* This distinction determines which variable appears in the rows (Variable 1) and the columns (Variable 2). Selecting the McNemar test will produce the test desired.

Computer-Generated Output

Table 5.3 presents the syntax commands and computer-generated output for the McNemar test. The syntax commands can be used in earlier versions of SPSS. They indicate that the sample selection has been restricted to only those children from the intervention group (Group = 0) because a filter has been placed on the data set (filter_$) (①) and that a McNemar test has been run on the restricted sample (②). The subcommand, */missing analysis,*

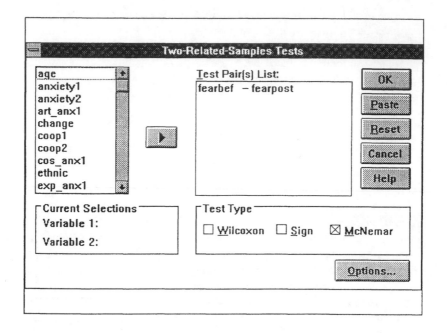

Figure 5.1. SPSS for Windows Computer Commands for the McNemar Test

indicates that we have selected the default option for the handling of missing values: Cases will be omitted on a test-by-test basis.

In the printout, we are presented with a table that is similar to Table 5.2 (③). The results indicate that 6 children who reported fearing the operation prior to the intervention no longer reported fear following the intervention (④). No children reported increased fear following the intervention (⑤).

In SPSS for Windows, to test whether this change is greater than what would be expected by chance, the chi-square approximation for the Mc-Nemar test is used if the sample size is sufficiently large (i.e., when the expected frequency, $(A + D)/2$, is at least 5 [Siegel & Castellan, 1988]. In our case, $(A + D)/2 = (0 + 6)/2 = 3$, which is smaller than the criterion. The printout in SPSS for Windows therefore represents the binomial test instead.

Notice that we are presented with a two-tailed p value (.0313) in the printout (⑥). Because our alternative hypothesis is directional, our test is

Table 5.3　Computer-Generated Printout From the McNemar Test

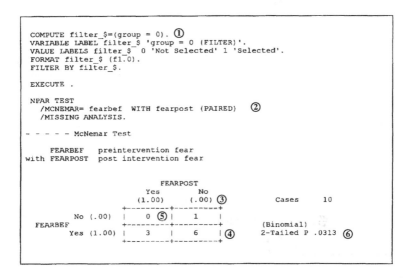

```
COMPUTE filter_$=(group = 0). ①
VARIABLE LABEL filter_$ 'group = 0 (FILTER)'.
VALUE LABELS filter_$  0 'Not Selected' 1 'Selected'.
FORMAT filter_$ (f1.0).
FILTER BY filter_$.

EXECUTE .

NPAR TEST
    /MCNEMAR= fearbef  WITH fearpost (PAIRED)  ②
    /MISSING ANALYSIS.

- - - - - McNemar Test

        FEARBEF      preintervention fear
with FEARPOST    post intervention fear


                       FEARPOST
                    Yes        No
                   (1.00)     (.00) ③       Cases      10
                 +---------+---------+
          No (.00) |  0 ⑤ |   1    |
FEARBEF          +---------+---------+        (Binomial)
         Yes (1.00) |  3   |   6    | ④      2-Tailed P  .0313 ⑥
                 +---------+---------+
```

one-tailed, and we need to divide the presented p value in half ($p = .0313/2$ = .016). This resulting value, .016, is less than our $\alpha = .05$. The null hypothesis is, therefore, rejected. Examining the direction of the results obtained in the computer printout (Table 5.3), we can conclude that, among the 10 children in the intervention group, the change in fear levels was in the direction of lowering fear.

Presentation of Results

Table 5.4 presents a suggested approach to presenting the results of the McNemar test. Percentages are omitted from the table to prevent confusion resulting from such a small sample. An alternative method would be to present the results in written format:

The results of the McNemar test indicate that the intervention was effective in changing fear levels in the direction of reduced fear from pre- to postintervention ($p = .016$).

Table 5.4 Suggested Presentation of McNemar Test Results

	Fear Following Intervention?		
	Yes	*No*	*Total*
Fear prior to intervention?			
Yes	3	6	9
No	0	1	1
Total	3	7	10

NOTE: One-tailed $p = .016$.

Advantages and Limitations of the McNemar Test

The McNemar test is especially useful because it can be used to examine change in nominal-level data. A disadvantage of this test is that it does not examine extent of change, only whether change has occurred. In the hypothetical example, it is not possible to ascertain the extent of change in the children's fears, only whether they changed, and if so, in what direction.

The McNemar test also does not allow for a comparison group (e.g., the control group) because this is a test using dependent observations. Feuer and Kessler (1989) have presented a two-sample situation for the McNemar test in which the marginal changes in a nominal-level variable for two independent cohorts (a control and an intervention cohort) are examined across two time periods. If the researcher wanted to compare two groups, two additional nonparametric approaches are possible. McNemar tests could be run on each of the independent groups independently and their results compared. Alternatively, a 2 x 2 contingency table representing group membership (intervention, control) by change classification (change in a positive direction, change in a negative direction) could be created. The Fisher's exact test (Chapter 7) could then be used to examine group by time interaction between the two dichotomous variables, group and direction of change.

Parametric and Nonparametric Alternatives to the McNemar Test

Because the McNemar test is used primarily with nominal-level data, there is no parametric counterpart to this test. If there are more than two periods of data collection (e.g., pretest, posttest, and followup), Cochran's Q test is recommended (see Chapter 6). If the data are continuous and meaningfully ranked, it would be advisable to use more sensitive non-parametric tests that use paired data, such as the sign test and the Wilcoxon test.

Examples From Published Research

Bustamante, E. A., & Levy, H. (1994). Sputum induction compared with broncho-alveolar lavage by Ballard catheter to diagnose Pheumocystis carinii pneumonia. *Chest, 105*, 816-822.

Henry, S. B. (1991). Effect of level of patient acuity on clinical decision making of critical care nurses with varying levels of knowledge and experience. *Heart and Lung: Journal of Critical Care, 20*, 478-485.

Snoey, E. R., Housset, B., Guyon, P., El Haddad, S., Valty, J., & Hericord, P. (1994). Analysis of emergency department interpretation of electrocardiograms. *Journal of Accident and Emergency Medicine, 11*, 149-153.

Wechsberg-Wendee, M., Cavanaugh, E. R., Dunteman, G. H., & Smith, F. J. (1994). Changing needle practices in community outreach and methadone treatment. *Evaluation and Program Planning, 17*, 371-379.

The Sign Test

The sign test is one of the oldest of all nonparametric tests, dating back to 1710 (Conover, 1980). It is called the sign test because the statistic is generated from data that have been reduced to +'s or −'s. The test can be used with paired data that are either dichotomous or continuous and that have been collected across a single sample or matched pairs. If the data are dichotomous, the two categories making up the variable need to have some rank order to their measurement (e.g., "success" vs. "failure" or "yes" vs. "no"), and the test reduces to the McNemar situation.

The sign test can be used in any situation in which the researcher can determine whether one of two paired or matched observations is "greater" or "less" than the other with regard to some identified attribute. The exact quantitative amount of the difference does not need to be determined for this test because the focus of the analysis is on the signs of the differences between each pair of variables.

An Appropriate Research Question for the Sign Test

The sign test has not been very popular in the research literature, possibly because it has been overshadowed by the more powerful Wilcoxon signed ranks test. Several examples from the literature, however, demonstrate the potential usefulness of the sign test. McAlindon and Smith (1994) used a sign test to assess the effectiveness of interactive video instruction in teaching the process of quality assurance to registered nurses (RNs) in the clinical setting. Hutchinson (1990) used a similar test to evaluate adolescent mothers' perceptions of newborn infants and their use of coping behaviors, and Katz and Sachs (1991) compared occupational therapy students and practitioners from the United States and Israel regarding concepts related to human performance.

In our hypothetical health care intervention study, suppose we were interested in examining the reduction in anxiety as assessed by the attending nursing staff from pretest to posttest within the group of 10 children who received the intervention. An example of a research question for which a sign test would be appropriate would be as follows:

Do the children in the intervention group reduce their anxiety from pretest to posttest?

Note that because the pretest and posttest anxiety measures that we are using are scales of ordinal measurement that range from 1 to 7, we cannot use the McNemar test. Had we known from the beginning of our study that we were planning to use the sign test for analysis of these data, it would have been unnecessary to collect data with so much detail. All we needed to do was ask the attending nursing staff whether or not the child had reduced ("–") or increased ("+") his or her anxiety from pretest to posttest.

Null and Alternative Hypotheses

Table 5.5 presents the null and alternative hypotheses for which a sign test would be appropriate given the research question outlined above. Note that the alternative hypothesis is directional; therefore, the test will be one-tailed.

Overview of the Procedure

To compute the sign test, the differences in the paired data are obtained and the direction of the differences ("+" or "–") recorded and summed. The test then takes the form of a binomial test (see Chapter 4 for calculations) in which the sum of the negative signs is compared to the sum of the positive signs, ignoring the instances of no differences. Depending on the

Table 5.5 Example of Null and Alternative Hypotheses Appropriate for the Sign Test

---

Null Hypothesis

H_0: The anxiety of the children who took part in the intervention will not change from pretest to posttest; that is, the number of children whose anxiety is reduced from pretest to posttest is the same as the number of children whose anxiety is increased.

Alternative Hypothesis

H_a: The anxiety of the children who took part in the intervention will decrease from pretest to posttest; that is, the number of children whose anxiety is reduced from pretest to posttest is greater than the number of children whose anxiety is increased.

---

direction stated in the alternative hypothesis, the null hypothesis of no difference between the number of positive and negative signs will be rejected if the probability of obtaining as extreme an occurrence of the obtained values is less than the prestated alpha level.

Either a one- or a two-tailed test may be used, depending on the wording of the alternative hypothesis. In a one-tailed test, the alternative hypothesis states which sign, positive or negative, will occur more frequently. A two-tailed alternative test merely states that there will be a difference in the number of positive and negative signs.

When the sample size is relatively large ($N > 25$ in SPSS for Windows), the normal approximation to the binomial distribution is used for the sign test. This distribution has a mean, μ_x, that is equal to np and a variance, σ_x^2, equal to npq. The value of the z statistic with a continuity correction for categorical data and $p = q = .5$ is given as follows:

$$z = \frac{x - \mu_x}{\sigma_x} = \frac{(x \pm 0.5) - np}{\sqrt{npq}} = \frac{(x \pm 0.5) - 0.5n}{0.5\sqrt{n}}$$

where

x = the number of $+$'s or $-$'s, depending on the stated direction of the alternative hypothesis

n = the number of paired observations that have been assigned a "$+$" or "$-$" value

When calculating the z statistic, $(x + 0.5)$ is used when $x < .5n$ and $(x - 0.5)$ is used when $x > .5n$. The calculated value of this z statistic is then compared to the critical value of the standard normal distribution at the prestated one- or two-tailed alpha level.

Critical Assumptions of the Sign Test

One of the advantages of the sign test is that there are not many assumptions attached to it. Unlike the paired t test, the sign test makes no assumptions regarding the form of the distribution of differences between the two variables being examined. The assumptions for this test are as follow.

1. The data to be analyzed may be dichotomous or continuous. For dichotomous data, there must be some order implied in the coding system (e.g., "0" and "1"). The data that we are examining, pre- and postintervention anxiety, have been measured on a 7-point Likert-type scale and are, therefore, at the ordinal level of measurement.

2. The randomly selected data are paired observations from a single sample, constructed either through matched pairs or through utilizing subjects as their own controls. The data from our hypothetical intervention study consist of two pre- and postintervention measures that have been conducted on the same sample of children and are, therefore, paired observations. The data are not, however, randomly selected.

Computer Commands

Figure 5.2 presents the SPSS for Windows dialogue box used to generate the sign test for pretest and posttest anxiety level for the 10 children who received the intervention. Note that the Wilcoxon signed ranks test also can be generated for the anxiety data from the same dialogue box by clicking on the appropriate test box. This dialogue box was opened by selecting the following items from the menu: *Statistics . . . Nonparametric Tests . . . 2 Related Samples.* As with the McNemar test, the sample was first reduced to include only the children in the experimental group using the commands *Data . . . Select Cases* and setting the condition that the children were from the intervention group (i.e., Group = 0).

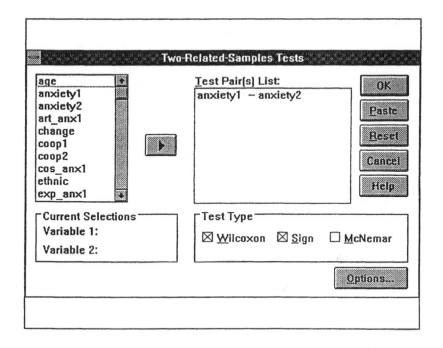

Figure 5.2. SPSS for Windows Computer Commands for the Sign and Wilcoxon Tests

Computer-Generated Output

Table 5.6 presents the syntax commands and computer-generated output for the sign test obtained from SPSS for Windows. The first bit of information that we are given is that the data have been filtered to include only those children who took part in the intervention (i.e., Group = 0) (①). Next, we are given the syntax commands for the sign test. Note that by opening the *Options* dialogue box, we have been able to select descriptive and quartile statistics (②) as well as the test-by-test default option for missing values (the command, */missing analysis*). The *Descriptives* command generates the mean, standard deviation, and minimum and maximum values for each group (③), and the median, or 50th percentile, is presented under *Quartiles* (④). Both the means and 50th percentile that were generated indicate that the children who took part in the intervention had lowered anxiety levels from Time 1 to Time 2.

Table 5.6 Computer-Generated Printout for the Sign Test

```
COMPUTE filter_$=(group = 0). ①
VARIABLE LABEL filter_$ 'group = 0 (FILTER)'.
VALUE LABELS filter_$ 0 'Not Selected' 1 'Selected'.
FORMAT filter_$ (f1.0).
FILTER BY filter_$.

EXECUTE .

NPAR TEST
   /SIGN= anxiety1  WITH anxiety2 (PAIRED)
   /STATISTICS DESCRIPTIVES QUARTILES  ②
   /MISSING ANALYSIS.


                N        Mean      Std Dev   Minimum   Maximum  ③

ANXIETY1       10     5.80000     1.39841     3.00      7.00
ANXIETY2       10     4.40000      .96609     3.00      6.00


                                (Median) ④
                        25th        50th        75th
                 N   Percentile  Percentile  Percentile

ANXIETY1       10      4.7500      6.0000      7.0000
ANXIETY2       10      3.7500      4.5000      5.0000


 - - - - - Sign Test

         ANXIETY1   preintervention anxiety
with ANXIETY2   post-test anxiety

         Cases

    ⑤    8  - Diffs (ANXIETY2 LT ANXIETY1)
         1  + Diffs (ANXIETY2 GT ANXIETY1)   (Binomial) ⑥
         1    Ties                      2-Tailed P = .0391 ⑦
         --
        10    Total
```

To determine whether these differences are sufficiently large to warrant rejecting the null hypothesis, we need to examine the number of negative differences ($n = 8$) (i.e., the number of children whose anxiety at posttest was lower than at pretest), the number of positive differences ($n = 1$) (i.e., those whose anxiety at posttest was greater than at pretest), and the number of ties (i.e., those for whom there was no change in anxiety scores) (⑤). Because the sample size was small ($N < 25$), a binomial test was undertaken (⑥). The resulting one-tailed p value of .02 (.0391/2 = .02) (⑦) is less than our stated level of alpha (.05); therefore, we will reject the null hypothesis

Table 5.7 Suggested Presentation of Sign Test Results

Anxiety Scores	N	Mean	Median	Standard Deviation	p*
Pretest	10	5.6	6.0	1.3	
					.02
Posttest	10	5.0	5.0	1.1	

*The calculated one-tailed p-value is for the sign test.

and conclude that there was a significant reduction in anxiety level from pretest to posttest among the children who took part in the intervention.

Presentation of Results

Table 5.7 is an example of how the results of the sign test might be reported. This type of table can be obtained in SPSS for Windows by highlighting the commands *Statistics . . . Custom Tables . . . Basic Tables* and choosing the summarizing statistics that are desired. Notice that the median (*Md*) is presented along with the mean and standard deviation. Given that the sign test is nonparametric, some authors might prefer to limit the presentation to only the medians. Because the sample size was small, the binomial distribution was used to evaluate the sign test. For that test, only the *p* value can be presented in the table. For larger sample sizes, the normal approximation to the binomial distribution is used. For that reason, the generated *z* statistic also could be presented.

The results from statistical analysis using the sign test could also be more easily presented in the text as follows:

The results of the sign test analysis indicated that the 10 children who took part in the clinical intervention significantly reduced their median anxiety levels from pretest (*Md* = 6.0) to posttest (*Md* = 5.0) (*p* < .05).

Advantages and Limitations of the Sign Test

The sign test is a versatile, simple, and easy-to-apply statistical test that can be used to determine whether one variable tends to be larger than another. It also can be used to test for trends in a series of ordinal measurements (Conover, 1980) or as a quick assessment of direction in an exploratory study. The disadvantage of this test is that it does not take into

account the order of magnitude of the differences between two paired
variables. When data are at least ordinal in level of measurement, the
Wilcoxon signed ranks test is preferred.

Parametric and Nonparametric Alternatives
to the Sign Test

The parametric alternative to the sign test is the paired t test. Both Siegel
and Castellan (1988) and Walsh (1946) report that the sign test is about
95% as efficient as the paired t test. Recall from Chapter 3 that *power
efficiency* refers to the sample size that is required for one test (e.g., the
sign test) to be as powerful as its rival (e.g., the paired t test) given the same
alpha level and that the assumptions of both tests have been met. A 95%
efficiency rating implies that, for small samples, only 20 cases are needed
for the sign test to achieve the same power as the paired t test with 19 cases
(i.e., N_2/N_1 [100%] = 19/20 [100%] = 95%). This suggests that the sign test
is especially useful for small sample sizes and in situations in which
meeting the assumptions of the robust paired t test either is not possible
(e.g., the data are nominal) or is questionable (e.g., a severely skewed
distribution with small sample sizes). A more powerful nonparametric
alternative to the sign test when the data are at least ordinal in level of
measurement is the Wilcoxon signed ranks test, which makes better use of
the quantitative differences between the paired observations.

Examples From Published Research

Hutchinson, S. W. (1990). Adolescent mothers' perceptions of newborn infants and
    the mothers' use of coping behaviors: A descriptive study. *Journal of the National
    Black Nurses' Association, 4*, 14-23.
Katz, N., & Sachs, D. (1991). Meaning ascribed to major professional concepts: A
    comparison of occupational therapy students and practitioners in the United
    States and Israel. *American Journal of Occupational Therapy, 45*, 137-145.
McAlindon, M. N., & Smith, G. R. (1994). Repurposing videodiscs for interactive
    video instruction: Teaching concepts of quality improvement. *Computers in
    Nursing, 12*, 46-56.

The Wilcoxon Signed Ranks Test

The reduction of data in the sign test to mere +'s or −'s results in the loss
of potentially important quantitative information: the *size* of the differences

between two paired variables. In our anxiety data, for example, no use is made by the sign test of the information that 5 of the 10 children reduced their anxiety by more than 2 points and that one child increased his anxiety by 3 points. By taking into account both the magnitude and the direction of changes, the Wilcoxon signed ranks test, which was developed by Wilcoxon (1945), produces a more sensitive statistical test. It is used with paired data that are measured on at least the ordinal scale and is especially effective when the sample size is small and the distribution of the data to be examined does not meet the assumptions of normality, as is required in the paired *t* test.

An Appropriate Research Question for the Wilcoxon Signed Ranks Test

The Wilcoxon signed ranks test has been used widely in the health care research literature. It is a very flexible test that can be used in a variety of situations with different sample sizes and few restrictions. The only requirements are that the data be continuous and be paired observations; that is, there are either pretest-posttest measures for a single sample or subjects who have been matched on certain criteria.

The Wilcoxon signed ranks test has been used frequently in the research literature to evaluate changes in attitudes on a variety of topics, such as changes over time in nursing students' attitudes toward Australian aborigines (Hayes, Quine, & Bush, 1994) and attitude changes in flight instructors (Alkov & Gaynor, 1991). It also has been particularly useful in evaluating the effectiveness of interventions, such as the effects of music on exercise repetitions in elderly women with osteoarthritis (Bernard, 1992), magnetic resonance imaging in patients with low-tension glaucoma (Stroman, Stewart, Golnik, Cure, & Olinger, 1995), and the results of benign breast biopsy on subsequent breast cancer detection practices of 238 women.

Numerous matched case/control studies have been conducted using the Wilcoxon signed ranks statistic. For example, Burt-McAliley, Eberhardt, and van Rijswijk (1994) examined the effects of skin barriers and adhesives on the incidence of peristomal skin irritation in 112 colostomy patients. Novack and colleagues (1994) used the test to evaluate focal intestinal perforation in matched pairs of extremely low-birth-weight infants. Giannini and Protas (1991) evaluated the aerobic capacity of 55 juvenile rheumatoid arthritis patients compared to a matched sample of healthy children, and Chaplin, Deitz, and Jaffe (1993) looked at motor performance in 14 children after traumatic brain injury compared with a sample of 14 normal

Table 5.8 Example of Null and Alternative Hypotheses Appropriate for the Wilcoxon Signed Ranks Test

---

Null Hypothesis

H_0:   The median anxiety scores of the children who took part in the intervention will not change from pretest to posttest (i.e., $Md_{pretest} = Md_{posttest}$).

Alternative Hypothesis

H_a: The median posttest anxiety score of the children who took part in the intervention will be lower than at pretest (i.e., $Md_{pretest} > Md_{posttest}$).

---

children matched for age and gender. There appear to be limitless possibilities for the application of the Wilcoxon signed ranks test.

In our hypothetical intervention study, we will continue to use the anxiety data that were collected on the children at pretest and then immediately following the clinical intervention. This will enable us to compare the results that we obtain from the Wilcoxon signed ranks test with those from the sign test. A research question similar to that of the sign test could, therefore, be asked:

> Do the children in the intervention group reduce their anxiety from pretest to posttest?

Null and Alternative Hypotheses

Table 5.8 presents an example of null and alternative hypotheses that would be appropriate for the Wilcoxon signed ranks test. Note that this nonparametric test examines the differences between medians, not means. Because our alternative hypothesis is directional (i.e., we are predicting a drop in anxiety level following our intervention), the test that will be undertaken is one-tailed. Our level of alpha for this test will remain the same as before ($\alpha = .05$).

Overview of the Procedure

To conduct the Wilcoxon signed ranks test, the differences between the paired data are calculated and the absolute values of these differences are recorded. Next, the absolute values of the differences between the two variables are ranked from lowest to highest. Finally, each rank is given a positive or negative sign depending on the sign of the original difference. The positive and negative ranks are then summed and averaged. Pairs that indicate no change are dropped from the analysis.

A z statistic is used to test the null hypothesis of no differences in the matched pairs. This z statistic takes the following form:

$$z = \frac{x - \mu}{\sigma} = \frac{T - [n\ (n+1)/4]}{\sqrt{n(n+1)\ (2n+1)/24}}$$

where

$T =$ the sum of the positive or negative ranks, depending on the proposed alternative hypothesis

$n =$ the number of positive and negative ranks, excluding ties.

If the null hypothesis is true, the sum of the positive ranks should be nearly equal to the sum of the negative ranks. If the differences in positive and negative ranks are sufficiently large, the null hypothesis is rejected. Either a one- or a two-tailed test is undertaken, depending on the wording of the alternative hypothesis.

Critical Assumptions of the Wilcoxon Signed Ranks Test

The assumptions of the Wilcoxon signed ranks test are fairly liberal.

1. The data are paired observations from a single randomly selected sample, constructed either through matched pairs or through utilizing subjects as their own controls. It is assumed either that the data being analyzed are test-retest measures of the same group of randomly selected subjects or that the data have been collected from subjects who have been paired on one or more variables. The data for our hypothetical study only partially meet this assumption. Although the anxiety data consist of Time 1 and Time 2 measures for the same sample of 10 children who took part in the intervention, our sample is a nonrandom sample of convenience.

2. The data to be analyzed must be continuous and at least ordinal in level of measurement, both within and between pairs of observations. This

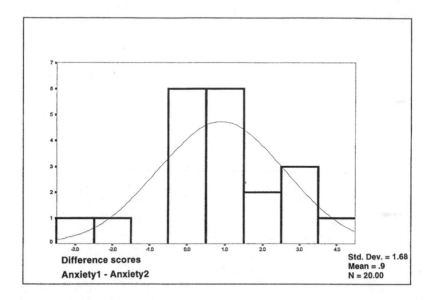

Figure 5.3. Histogram of Difference Scores: Anxiety1 – Anxiety2

assumption means that not only must the variables themselves be at least ordinal in level of measurement, but also the generated values of the difference scores must be continuous. In fact, Daniel (1990) indicates that these differences should be measured on at least an interval scale.

The anxiety data from our hypothetical intervention study consist of two pretest and posttest Likert-type scale measurements (1 = *not at all anxious* and 7 = *extremely anxious*). Both of these scales and their difference scores are continuous and at least ordinal in level of measurement.

3. There is symmetry of the difference scores about the true median for the population. This assumption implies that, if it were possible to view the distribution of the difference scores in the population, the distribution of these difference scores would be symmetric (though not necessarily normal) about the population median (Daniel, 1990). One approach to assessing this third assumption might be to plot the difference scores for the sample to assess their symmetry. Figure 5.3 presents the plot that was generated for the DIFF12 variable that was created by subtracting the children's Anxiety1 scores from their Anxiety2 scores using the *Transform . . . Compute* commands. This histogram was obtained by opening the *Statistics . . . Summarize . . . Frequencies* dialogue box and selecting histograms from the *Charts* option. The histogram indicates that although the data for the 20 children are not completely symmetric, they are not badly skewed. We could conclude, therefore, that we approach meeting this assumption.

Computer Commands

The SPSS for Windows dialogue box that was used to generate the sign test (Figure 5.2) also produces the Wilcoxon signed ranks test. The Wilcoxon signed ranks test was obtained by clicking on the *Wilcoxon* box under *Test Type*. As with the sign test, summary statistics information (e.g., descriptive statistics and quartiles) can be obtained by opening the *Options* menu.

Computer-Generated Output

Non para — 2 related Samples

Table 5.9 presents the syntax commands and computer-generated output from SPSS for Windows for the Wilcoxon signed ranks test. As with the previous tests, because we are interested only in the results for the 10 children in the intervention group, the *Select Cases* command obtained from the *Data* menu is operative (①). The SPSS for Windows syntax commands for the Wilcoxon Signed Ranks test are then presented (②). These commands can be used with earlier versions of SPSS.

The computer-generated printout for the Wilcoxon signed ranks test indicates that 8 children had a negative rank (their anxiety scores at posttest were less than those at pretest), 1 child increased his anxiety from pretest to posttest, and 1 child did not alter his or her anxiety score (③). SPSS for Windows then presents the mean or average rank for the difference scores of this group. It is interesting that the average change in rank for the children who decreased their anxiety was *smaller* (4.75) than that for the single child whose anxiety *increased* from pretest to posttest (7.00) (④). This would lead us to think that we are getting opposite results from what we predicted.

The z statistic that was generated from these data (−1.8363) (⑤) was obtained by the formula for the z statistic for the Wilcoxon Signed Ranks test as follows:

$$z = \frac{x-\mu}{\sigma} = \frac{T-[n(n+1)/4]}{\sqrt{n(n+1)(2n+1)/24}} = \frac{7-[9(9+1)/4]}{\sqrt{(9+1)(2(9)+1)/24}} = \frac{7-22.5}{8.441} = -1.8363.$$

Table 5.9 Computer-Generated Printout From SPSS for Windows for the
Wilcoxon Signed Ranks Test

```
COMPUTE filter_$=(group = 0).  ①
VARIABLE LABEL filter_$ 'group = 0 (FILTER)'.
VALUE LABELS filter_$  0 'Not Selected' 1 'Selected'.
FORMAT filter_$ (f1.0).
FILTER BY filter_$.

EXECUTE .

NPAR TEST
   /WILCOXON=anxiety1  WITH anxiety2 (PAIRED)  ②
   /MISSING ANALYSIS.

- - - - - Wilcoxon Matched-Pairs Signed-Ranks Test

        ANXIETY1   preintervention anxiety
with ANXIETY2   post-test anxiety

     Mean Rank      Cases
④     4.75          8  - Ranks (ANXIETY2 LT ANXIETY1)  ③
      7.00          1  + Ranks (ANXIETY2 GT ANXIETY1)
                    1    Ties  (ANXIETY2 EQ ANXIETY1)
                    --
                   10    Total

      Z =   -1.8363  ⑤            2-Tailed P =  .0663  ⑥
```

The T value that has been used by SPSS for Windows is the *sum* of the positive differences ([7][1] = 7). Had we used the sum of the negative differences ([8][4.75] = 38), we would have arrived at a positive sign for the z statistic. The two-tailed p value for this z statistic is .0663 (⑥). Based on this information, what is our decision regarding the null hypothesis that the medians are equal?

If our alternative hypothesis test had been nondirectional, we would have had to conclude that we could not reject the null hypothesis because $\alpha = .05$ is less than the two-tailed p value (.0663). Our alternative hypothesis, however, is directional, in that we stated that the children would have lower median anxiety scores at pretest than at posttest (Table 5.8). The median posttest value for the anxiety scores (5.0) *is* lower than the median pretest value (6.0) (Table 5.7). The mean rank of the anxiety scores that are presented in the computer printout, however, indicates a *higher* mean rank for the child whose anxiety scores increased from pre- to posttest (Table 5.9, ④). Is this information contradictory? What is our decision? Should we reject or fail to reject the null hypothesis?

It is apparent from examining both the median scores and the sum of the ranks that the positive and negative sums are in the direction predicted. That is, the *sum* of the positive ranks for the 8 children whose anxiety scores decreased (Σranks = $n[\overline{X}_{ranks}]$ = 8[4.75] = 38) is greater than the sum of the ranks for the one child whose anxiety increased from Time 1 to Time 2 (1[7] = 7). The average change for the larger group of 8 children, however, is not as great as that for the single child. The point is that the *mean rank* that is presented in the SPSS for Windows printout can be somewhat misleading and could, potentially, result in a misinterpretation of the data if the researcher is not careful. You should always examine the medians as well as the mean ranks to be certain that the results you are reporting make intuitive sense.

To determine whether this difference in sums is large enough to reject the null hypothesis, we need to compare our one-tailed α = .05 to half the presented two-tailed p value (.0663/2 = .0332). Because this one-tailed p value, .0332, is less than our α = .05 and our medians are in the direction predicted, we can reject the null hypothesis of equal medians. Based on our examination of the medians from Table 5.7, our conclusion is that the intervention group significantly reduced its anxiety from pretest to posttest.

Note that this generated one-tailed p value for the Wilcoxon test, .0332, is greater than the p value that was generated for the sign test, .02 (Table 5.6). The reason for this discrepancy is that the Wilcoxon signed ranks test is picking up the *size* of the negative and positive ranks, not just noting the direction of differences. The significance of the Wilcoxon statistic has been influenced by the fact that the one child who did increase his or her anxiety from pretest to posttest did so substantially.

Presentation of Results

The Wilcoxon signed ranks test can be presented in tabular form in much the same way as the sign test (Table 5.7). The only change that would be required would be the size of the p value (.03) and the type of test reported (Wilcoxon signed ranks test). The researcher may also wish to present median changes rather than the medians themselves. The results could also be presented in the text as follows:

The results of the Wilcoxon signed ranks test indicated that the 10 children who took part in the clinical intervention significantly reduced their median anxiety levels from pretest (*Md* = 6.0) to posttest (*Md* = 5.0) (p = .03).

Advantages and Limitations of the Wilcoxon Signed Ranks Test

The Wilcoxon signed ranks test is very easy to apply and has an advantage over the sign test in that it uses more of the information provided by the data. Although not a problem with the test per se, a potential disadvantage to the computer printout for this statistic in SPSS for Windows is that the *mean* ranks of the positive and negative ranks could be misinterpreted because *mean* ranks are not necessarily the same as the *sum* of the ranks used to compute this statistic.

Parametric and Nonparametric Alternatives to the Wilcoxon Signed Ranks Test

The parametric alternative to the Wilcoxon signed ranks test is the *t* test for related samples or paired *t* test. This is the statistical test of choice if the data are found to be normally distributed. The paired *t* test, however, has been found to be unsatisfactory when the distributions of the variables being considered have heavy tails (Wilcox, 1992). Blair and Higgins (1985) used Monte Carlo methods to assess the relative power of the paired *t* test and the Wilcoxon signed ranks test under 10 different distributional shapes. They report that, when the data were not normally distributed, the Wilcoxon was more often the more powerful test and that the magnitude of the Wilcoxon's power advantage over the paired *t* test often increased with the sample size.

The nonparametric alternatives to this test are the sign and binomial tests. As indicated above, these tests are not as sensitive as the Wilcoxon signed ranks test but could be used if that test's assumptions are not sufficiently met.

Examples From Published Research

Alkov, R. A., & Gaynor, J. A. (1991). Attitude changes in Navy/Marine flight instructors following an air crew coordination training course. *International Journal of Aviation Psychology, 1,* 245-253.

Bernard, A. (1992). The use of music as purposeful activity: A preliminary investigation. *Physical and Occupational Therapy in Geriatrics, 10,* 35-45.

Burt-McAliley, D., Eberhardt, D., & van Rijswijk, L. (1994). Clinical study: Peristomal skin irritation in colostomy patients. *Ostomy Wound Management, 40,* 28-37.

Chaplin, D., Deitz, J., & Jaffe, K. M. (1993). Motor performance in children after traumatic brain injury. *Archives of Physical Medicine and Rehabilitation, 74,* 161-164.

Giannini, M. J., & Protas, E. J. (1991). Aerobic capacity in juvenile rheumatoid arthritis patients and healthy children. *Arthritis Care and Research, 4,* 131-135.

Hayes, L., Quine, S., & Bush, J. (1994). Attitude change amongst nursing students towards Australian Aborigines. *International Journal of Nursing Studies, 31,* 67-76.

Novack, C. M., Waffarn, F., Sills, J. H., Pousti, T. J., Warden, M. J., & Cunningham, M. D. (1994). Focal intestinal perforation in the extremely-low-birth-weight infant. *Journal of Perinatology, 14,* 450-453.

Stroman, G. A., Stewart, W. C., Golnik, K. C., Cure, J. K., & Olinger, R. E. (1995). Magnetic resonance imaging in patients with low-tension glaucoma. *Archives of Ophthalmology, 113,* 168-172.

Discussion

In this chapter, we have examined three nonparametric statistical techniques that could be used when data collected from a single sample are paired through using subjects either as their own controls (e.g., pretest and posttest measures) or as matched pairs (e.g. husband-wife pairs). The McNemar test is used when the pretest-posttest data being examined are dichotomous. The sign and Wilcoxon tests require that the distribution of the variable being considered be continuous. Although the sign test can be used with ordered dichotomous data, the Wilcoxon signed ranks test assumes at least an ordinal level of measurement.

In Chapter 6, several nonparametric tests that can be used when data have been collected over more than two time periods will be presented. These tests include Cochran's Q test, which is used with dichotomous data, and the Friedman test, which is used with continuous data.

6 Repeated Measures for More Than Two Time Periods or Matched Conditions

- **Cochran's Q test**
- **Friedman test**

In Chapter 5, we examined tests that are useful for examining data for matched pairs or a single sample with "before and after" observations. In clinical research, we are sometimes interested in extending our data collection to include repeated observations across more than two time periods, such as preintervention, postintervention, and 6 months follow-up. Because these data are collected for more than two time periods, the McNemar, sign, and Wilcoxon signed ranks tests are no longer suitable choices for statistical analysis.

It may also be that we have blocks of subjects who have been matched on a *nuisance* factor that is thought to have an unwanted effect on the dependent variable of interest. This type of design is called a *randomized block design* (Wampold & Drew, 1990), in that blocks of matched subjects are formed such that the number of matched subjects in each block is equal to the number of levels of the independent variable. For example, subjects might be matched or *blocked* according to diagnosis. A subject from each diagnosis block is then randomly assigned to each level of the independent variable. The intent of this matching is to obtain an equal number of patients with given diagnoses in each group, thus reducing the amount of error variance that could be attributed to this factor. For this method to be effective, it is important that the researcher be certain that this blocking factor (e.g., patient diagnosis) does indeed have a statistically significant effect on the dependent variable (e.g., postoperative adjustment).

In this chapter, we will examine two nonparametric tests that can be used for multiple or repeated observations on single or matched samples: Cochran's Q test and the Friedman test. Cochran's Q test is used with dichotomous data, whereas the Friedman test is used with multiple observations of continuous data that are at least at the ordinal level of measurement.

Cochran's Q Test

As indicated in Chapter 5, the McNemar test is used to evaluate change in a dichotomous variable across two time periods. Cochran's Q test extends this test to examine change in a dichotomous variable across more than two observations. It is a particularly appropriate test when subjects are used as their own controls and the dichotomous outcome variable is measured across multiple time periods or under several types of conditions. For example, using a randomized block design, a group of subjects may be exposed to more than one intervention. If the treatment outcome can be classified into one of two categories (e.g., success or failure), Cochran's Q test would be appropriate for examining whether the proportion of cases in a particular category is the same across all conditions or time periods.

An Appropriate Research Question for Cochran's Q Test

Cochran's Q test has been used in a wide variety of statistical analyses in health care research. Lyman (1982) reported the use of Cochran's Q test to examine food preference when experiencing certain emotional states in a group of 100 undergraduates. McAlindon and Smith (1994) used Cochran's Q test as a test of reliability in their development of an interactive video instructional program for RNs. Shure and Astarita (1983) used Cochran's Q test to determine the optimal number of biopsy specimens that should be taken of an endobronchial mass lesion to verify the diagnosis of bronchogenic carcinoma. In dentistry, Sewerin (1994) used Cochran's Q test to evaluate agreement between observers regarding the resolution quality of the Ultra-Vision screen compared with a second system.

Siegel and Castellan (1988) have also illustrated the use of Cochran's Q to evaluate which items of a test are more difficult to answer correctly than others for a given group of respondents. In this example, the samples of dichotomous responses are the correct or incorrect answers to each item. These are related samples because it is assumed that the same group of subjects has answered all the test items.

In our hypothetical intervention example, the children from the intervention group were asked prior to and following the intervention to identify whether or not they were fearful of their upcoming operations. The McNemar test from Chapter 5 evaluated the direction of change in children's fears from pretest to posttest. Suppose a third point of data collection took place shortly before the administration of the preoperative medication. Because there are now three points of data collection, the McNemar test can no longer be used. Cochran's Q test, however, would be appropriate. The following research question could then be posed:

Is the proportion of children who indicate fear (yes, no) of their upcoming operation the same across all three data collection periods?

Null and Alternative Hypotheses

Table 6.1 presents an example of null and alternative hypotheses for which a Cochran's Q test would be appropriate. Note that the alternative hypothesis is not directional. Like the F test in ANOVA, Cochran's Q test determines whether or not there is a significant difference in the proportion of specific responses across the three conditions. Post hoc tests using the McNemar test with a Bonferroni correction for increased Type I error are necessary to determine where the differences lie.

Overview of the Procedure

To undertake a Cochran's Q test, the data are first cast into a two-way table consisting of N rows and k columns, where N represents the number of subjects and k represents the number of conditions or time periods (Siegel & Castellan, 1988). Table 6.2 presents the frequency data generated for the 10 children in our hypothetical intervention study.

Next, the data are evaluated to determine whether the proportion of subjects who indicate a preference for a particular dichotomous response (e.g., "yes" = 1) is similar across all k conditions. The statistic that is used to evaluate these proportions is as follows (Cochran, 1954):

$$Q = \frac{(k-1)\,[k\sum\limits_{j=1}^{k} G_j^2 - (\sum\limits_{j=1}^{k} G_j)^2]}{k\sum\limits_{i=1}^{N} L_i - \sum\limits_{i=1}^{N} L_i^2}$$

Table 6.1 Example of Null and Alternative Hypotheses Appropriate for Cochran's Q Test

---

Null Hypothesis

H_0: The proportion of "yes" responses with regard to their fearfulness concerning their upcoming operation will be the same across all three time periods for those children who took part in the intervention.

Alternative Hypothesis

H_a: The proportion of "yes" responses will differ for the children taking part in the intervention depending on which of the three time periods is being considered.

---

where

k = the number of conditions or time periods

N = the number of subjects

G_j = the total number of 1s in the jth column

L_i = the total number of 1s in the ith row

This Q statistic is distributed approximately as a chi-square (χ^2) with $df = k - 1$ if the size of the sample is greater than 4 and the calculated value, $(N)(k)$, is greater than 24. For smaller sample sizes, Patil (1975) offers exact distributions of Q, but as Siegel and Castellan (1988) point out, Cochran's Q quickly approximates the χ^2 distribution. The null hypothesis of equality of proportions of 1s across all levels of k will be rejected if the calculated value of Q exceeds the critical value of the χ^2 with $df = k - 1$ at the prespecified level of alpha (e.g., $\alpha = .05$). If there are only two time periods, Cochran's Q is nothing more than the McNemar test.

The Q statistic for the data from our hypothetical study will be distributed approximately as a χ^2 with $df = 2$ because our sample size is greater than 4 ($N = 10$), and $(N)(k) > 24$, that is, $(10)(3) > 24$. Calculating Cochran's Q from the data presented in Table 6.2, we obtain the following value:

Table 6.2 Frequencies of Intervention Group Children's Responses to Fear (1 = Yes, 0 = No)

| Subject | Children's Fear (k = 3) | | | L_i (Row 1's) |
	Preintervention	Postintervention	Preoperative	
1	1	0	1	2
2	1	0	0	1
3	1	1	1	3
4	0	0	0	0
5	1	0	1	2
6	1	0	1	2
7	1	1	1	3
8	1	0	0	1
9	1	1	1	3
10	1	0	0	1
G_j (column 1's)	9	3	6	18

$$Q = \frac{(k-1)\ [k\sum_{j+1}^{k} G_j^2 - (\sum_{j+1}^{k} G_j)^2]}{k\sum_{i+1}^{N} L_i - \sum_{i+1}^{N} L_i^2}$$

$$= \frac{(3-1)\ [3(9^2+3^2+6^{2)}-(18^2)]}{3(2+1+3+2+2+3+1+3+1)-(2^2+1^2+3^2+2^2+2^2+3^2+1^2+3^2+1^2)}$$

$$= \frac{2\ (378-324)}{3\ (18)-42} = \frac{108}{12} = 9.$$

This generated Q value, 9, is greater than the critical value of the χ^2 at $df = 2$ and $\alpha = .05$: $\chi^2 = 5.99$. The null hypothesis of equality of proportions of "yes" responses across all three time periods therefore will be rejected. The implications of this finding and its significance level will be addressed when we examine the computer printout.

Critical Assumptions of the Cochran's Q Test

Cochran's Q test shares many of the assumptions of the McNemar test but extends these assumptions to more than two conditions or time periods.

1. The variables being assessed across all time periods are dichotomous with levels that have been assigned similar values (e.g., 0s and 1s). This assumption means that the same dichotomous variables have been measured across multiple conditions or time periods. The researcher also needs to make certain that the levels that are coded with 0s and 1s share the same meaning across all time periods. In our hypothetical intervention example, the children were asked to assess their fear (yes, no) of the upcoming operation across three time periods—preintervention, postintervention, and preoperatively. Their "yes" or "no" answers were coded similarly each time (1 = yes, 0 = no); therefore, these data meet this assumption.

2. The data being examined represent frequencies, not scores. The data being examined are the frequencies of occurrence of a particular event (e.g., the number of times in a particular time period that a "yes" or a "1" occurred). These data do not represent scores on a particular test. Our hypothetical data, children's responses to a question concerning their fear (yes, no), meet this assumption.

3. The dichotomous measures are multiple observations of the same subjects or matched samples. This assumption implies that either (1) the dichotomous data have been collected on the same subjects across more than two time periods, (2) the same group of subjects have been exposed to multiple conditions, or (3) N sets of three or more subjects have been matched on certain characteristics and that, within each set, each matched subject has been randomly assigned to one of the designated conditions. In our hypothetical intervention study, the children who took part in the intervention were asked the same question at three different points in time: prior to the intervention (*Fearbef*), following the intervention (*Fearpost*), and just prior to administration of the preoperative medication (*Fear_op*). Our data, therefore, meet this assumption.

4. The levels of the categorical variables are mutually exclusive. This assumption means that a subject can be assigned to only one level (0 or 1) of the dichotomous variable at each of the time periods or conditions. The children in the intervention group could be assigned only a "1" or a "0,"

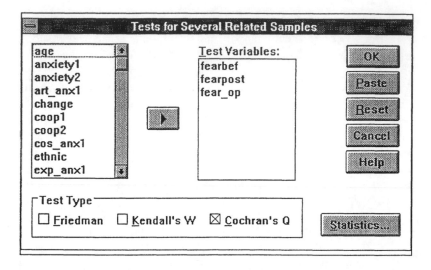

Figure 6.1. SPSS for Windows Computer Commands for Cochran's Q Test

depending on whether or not they expressed fear concerning their impending operation. The levels of these dichotomous variables (*Fearbef*, *Fearpost*, and *Fear_op*) therefore are mutually exclusive.

Computer Commands

Figure 6.1 presents the SPSS for Windows commands used to generate Cochran's Q test. This dialogue box was opened by selecting the following items from the menu: *Statistics . . . Nonparametric Tests . . . k Related Samples.* As with the McNemar test from Chapter 5, the analysis will be restricted to only those 10 children who took part in the intervention. This restriction is obtained by opening the *Data . . . Select Cases* dialogue box and selecting only those cases that meet the condition Group = 0. Notice that the third variable, *fear_op* (fear prior to preoperative medication), has been added to the analysis and the test type, Cochran's Q, has been selected.

Computer-Generated Output

The SPSS for Windows syntax and computer-generated output are presented in Table 6.3. A filter variable (*filter_$*) has been created as a result

Table 6.3 SPSS for Windows Computer-Generated Printout for Cochran's
Q Test

```
COMPUTE filter_$=(group = 0).  ①
VARIABLE LABEL filter_$ 'group = 0 (FILTER)'.
VALUE LABELS filter_$ 0 'Not Selected' 1 'Selected'.
FORMAT filter_$ (f1.0).
FILTER BY filter_$.

EXECUTE .

NPAR TESTS
  /COCHRAN = fearbef fearpost fear_op  ②
  /MISSING LISTWISE.


- - - - - Cochran Q Test
     Cases
  = 1.00  =   .00    Variable  ③

       9      1    FEARBEF     preintervention fear
       3      7    FEARPOST    post intervention fear
       6      4    FEAR_OP     fear -preoperative medication  ④

        Cases           Cochran Q          D.F.  Significance
         10             9.0000  ⑤           2      .0111   ⑥
```

of the *Select Cases* command that excludes the children from the control
group from the analysis. The syntax commands therefore indicate that only
the data for the 10 children who took part in the intervention (Group = 0)
will be examined (①). The syntax commands also specify that Cochran's
Q test will be generated (②). The command "/MISSING LISTWISE"
means that cases that are missing from any of the listed variables will be
excluded from all analyses.

The frequencies of children who reported fear (1s) and no fear (0s) for
the three variables, preintervention, postintervention, and premedication
fear, are then presented (③). Note that although the number of children
who expressed fear of upcoming operations declined from preintervention
($n = 9$) to postintervention ($n = 3$), the frequency of children who expressed
fear increased as the operation became more imminent ($n = 6$) (④).

The value for the Cochran's Q test is 9.00 (⑤), which, thankfully, is what
we had calculated earlier. The generated p value for the Q statistic is .0111
(⑥), which is less than our $\alpha = .05$; therefore, the null hypothesis is rejected

Table 6.4 Suggested Presentation of Cochran's Q Test Results

	Fear of Operation?		*Cochran's* Q	
Time Period	*Yes*	*No*	χ^2	p
Pretest*	9	1		
Posttest	3	7	9.00	.01
Follow-up	6	4		

*Significantly different from posttest, $p = .015$.

and we can conclude that the fear levels of the children in the intervention group are significantly different across the three time periods.

To determine where the differences lie, it would be necessary to undertake several McNemar tests to compare each time period with the others. Because of the risk of increased Type I error, the alpha for these post hoc tests should be adjusted using Bonferroni's inequality (adjusted $\alpha' = \alpha/k$, where k = the number of post hoc tests to be performed and α is the original alpha level). In our hypothetical intervention study, we have three posttests: Time 1 versus Time 2, Time 1 versus Time 3, and Time 2 versus Time 3. Our adjusted α' therefore is .017 (i.e., $\alpha/k = .05/3 = .017$). Had we performed these tests (see Chapter 5), we would have found that the only two time periods that were significantly different from each other were pre- and postintervention fear (*Fearbef* and *Fearpost*).

Presentation of Results

Table 6.4 presents a format that might be used to present the results of the Cochran's Q test. This material could also be presented in the text as follows:

The results of Cochran's Q test indicate that there was a significant change in the fear levels in the children from the intervention group over the three time periods ($p = .01$). To determine where the differences lay, post hoc tests were undertaken using the McNemar test with a Bonferroni correction (one-tailed $\alpha = .017$) to accommodate for the increased risk of Type I error. These analyses indicated that the significant decreases in the intervention group children's reported fear occurred from pretest to posttest ($p = .015$). No other significant changes between time periods were obtained.

Advantages and Limitations of Cochran's Q Test

Cochran's Q test has the advantage of examining change in categorical data over multiple observations. A disadvantage to this test is that it does not evaluate the extent of change, only whether a change has occurred. It is also not very accurate when the sample size is small, nor does it accommodate a control group, because this is a test for use with dependent observations. A somewhat unsatisfactory solution is to run a Cochran's Q test on each of the two groups independently and to compare the results. Unfortunately, this approach does not allow the researcher to assess group by time interaction.

Parametric and Nonparametric Alternatives to Cochran's Q Test

Because Cochran's Q test is intended for use with dichotomous data, it has no parametric equivalent. If the data to be analyzed are continuous, the Friedman test is preferable to Cochran's Q test, especially when the sample size is small and the data are ordered (Siegel & Castellan, 1988). Myers, DiCecco, White, and Borden (1982) reported that Cochran's Q test had problems with sample sizes of less than 16 but gave honest Type I error rates for larger samples, even under conditions of extreme heterogeneity of covariance.

Examples From Published Research

Lyman, B. (1982). The nutritional values and food group characteristics of foods preferred during various emotions. *Journal of Psychology, 112*, 121-127.

McAlindon, M .N., & Smith, G. R. (1994). Repurposing videodiscs for interactive video instruction: Teaching concepts of quality improvement. *Computers in Nursing, 12*, 46-56.

Sewerin, I. P. (1994). Clinical testing of the Ultra-Vision screen-film system for maxillofacial radiography. *Journal of Oral Surgery, Oral Medicine & Oral Pathology, 77*, 302-307.

Shure, D., & Astarita, R. W. (1983). Bronchogenic carcinoma presenting as an endobronchial mass. *Chest, 83*, 865-867.

The Friedman Test

In Chapter 5, the Wilcoxon signed ranks test was presented as a useful technique for analyzing continuous data that have been collected on a single sample across two time periods or conditions. It also could be used when subject pairs are matched and randomly assigned as pairs to an experimental or control group. The Friedman test extends the Wilcoxon signed ranks test to include (1) more than two time periods of data collection or conditions, or (2) groups of three or more matched subjects, with a subject from each group being randomly assigned to one of the three or more conditions.

The Friedman test examines the ranks of the data generated during each time period or condition to determine whether the variables share the same underlying continuous distribution. This nonparametric test is analogous to the parametric repeated-measures ANOVA without a comparison group.

An Appropriate Research Question
for the Friedman Test

There are numerous examples from the health care research literature that illustrate the variety of situations to which the Friedman test can be applied. Philip, Ayyangar, Vanderbilt, and Gaebler-Spira (1994) used the Friedman test to evaluate the effects of rehabilitation on the functional outcomes of 30 children after treatment for primary brain tumors. McCain (1992) used the same test to assess the effectiveness of three interventions designed to facilitate inactive awake states in 20 preterm infants. Blegen and colleagues (1992) also made use of this test to examine 341 staff nurses' preferences for selected head nurse work performance recognition behaviors. Bertoti (1988) used the Friedman test to measure postural changes in 11 children with spastic cerebral palsy after participation in a therapeutic horseback riding program, and Crawford and McIvor (1985) used the same test to investigate the impact of group psychotherapy on changes in the psychological adjustment of matched sets of patients with a primary diagnosis of MS.

In our hypothetical intervention example, suppose we had collected information concerning the 10 children's anxiety levels not only at pretest and immediately following the anxiety reduction intervention but also just prior to the administration of the preoperative medication. An example of a research question that could be answered with the Friedman test is as follows:

What are the differences in the anxiety levels of the 10 children who took part in the intervention across the three time periods (i.e., preintervention, postintervention, and just prior to the administration of the preoperative medication)?

Table 6.5 Example of Null and Alternative Hypotheses for Which a Friedman Test Would Be Appropriate

---

Null Hypothesis

H_0: There will be no differences among the median anxiety scores at preintervention, at postintervention, and at preoperative medication for the 10 children who took part in the intervention.

Alternative Hypothesis

H_a: There will be at least one difference among the median anxiety scores at preintervention, at postintervention, and at preoperative medication for the 10 children who took part in the intervention.

Null and Alternative Hypotheses

Table 6.5 presents an example of null and alternative hypotheses, generated from the research question presented earlier, that would be appropriate for a Friedman test. Note that because this test is nonparametric, the focus of attention in the hypotheses is on the medians, not the means. In addition, the alternative hypothesis presented in Table 6.5 is nondirectional. If the alternative hypothesis were directional, planned comparisons would be a more powerful approach to analyzing the data. Whether they are parametric or nonparametric, overall tests of significance followed by post hoc tests are used primarily when the researcher is not certain of direction and, therefore, prefers to explore rather than predict outcomes.

Overview of the Procedure

To undertake a Friedman test, the data are first cast into a two-way table with N rows and k columns. As with Cochran's Q test, N represents the number of subjects or matched sets of subjects and k represents the number of conditions or data collection periods. For example, if there were 6 sets of 4 matched subjects that were going to be randomly assigned to one of four conditions, $N = 6$ and $k = 4$. In our hypothetical intervention example, because we have 10 subjects whose anxiety levels were measured at three time periods, $N = 10$ and $k = 3$. Table 6.6 presents the anxiety data for the 10 children in the intervention group across the three periods of time.

Note that the data in Table 6.6 for each row (i.e., for each matched set or subject) have been ranked from lowest to highest and their columns

Table 6.6   Frequencies of Anxiety Data for the 10 Children in the Intervention Group

	Preintervention		Postintervention		Preoperative	
ID	Score	Rank	Score	Rank	Score	Rank
1	7	3	5	1	6	2
2	4	2	4	2	4	2
3	6	3	5	1.5	5	1.5
4	3	1	6	3	4	2
5	6	3	3	1	4	2
6	7	3	3	1	6	2
7	6	3	5	1.5	5	1.5
8	7	3	5	1	6	2
9	5	3	4	1.5	4	1.5
10	7	3	5	1	6	2
Sum (R_j)		27		14.5		18.5

summed. If the null hypothesis of no differences between the conditions or time periods is true, the sum of the ranks for each column should be no different from that which would be expected by chance (i.e., $N[k + 1]/2$) (Siegel & Castellan, 1988). In our example, if there are no differences between the children's preintervention, postintervention, and premedication anxiety, the sum of the ranks for each of these three time periods should be equal to 20 (10[3 + 1]/2 = 20). If the null hypothesis is not true, then the sum of the ranks would vary from column to column. Our sums in Table 6.6—27, 14.5, and 18.5—are not equal. Are these differences sufficiently large to reject the null hypothesis?

The Friedman test examines the rank totals of each time period or condition to determine the extent to which these totals differ from their expected sums using the following formula (Siegel & Castellan, 1988):

$$F_r = \frac{12}{N k (k+1)} \left[\sum_{j-1}^{k} R_j^2 \right] - [3 N (k+1)]$$

where

$$R_j = \text{the sum of the ranks for column } j$$

N = the number of subjects

k = the number of time periods or conditions.

If the sample size (N) and number of conditions (k) are sufficiently large, this statistic is distributed as a χ^2 with $df = k - 1$. Siegel and Castellan (1988) also point out that when there are many ties among the ranks for a given row, F_r needs to be adjusted to account for changes in its sampling distribution. These authors present the somewhat cumbersome formula, which is rather tedious to calculate by hand. It also does not appear that the Friedman test in SPSS for Windows makes this adjustment.

Like the overall F test in ANOVA, if the obtained value of the Friedman statistic is significant, the researcher can conclude that at least one condition or time period is significantly different from another. To determine where the differences lie, post hoc tests need to be undertaken.

For our data in Table 6.6, we would obtain the following F_r:

$$F_r = \frac{12}{N\,k\,(k+1)} \left[\sum_{j-1}^{k} R_j^2 \right] - [3\,N\,(k+1)]$$

$$= \frac{12}{10\,(3)\,(3+1)} \, [27^2 + 14.5^2 + 18.5^2] - [3(10)(3+1)]$$

$$= \frac{12}{120} \, [1281.5] - [120]$$

$$= 128.15 - 120$$

$$= 8.15.$$

If the assumptions of the Friedman test have been met, our critical χ^2 value with $df = 3 - 1 = 2$ and $\alpha = .05$ is 5.99, which is smaller than our obtained F_r of 8.15. This means that the null hypothesis will be rejected. The meaning of this decision and the significance level for this F_r will be addressed when we examine the computer printout.

Critical Assumptions of the Friedman Test

The Friedman test shares the assumptions of the Wilcoxon signed ranks test but extends these assumptions to include more than two conditions or periods of data collection.

1. The data to be analyzed are continuous and at least at the ordinal level of measurement. Our data, the measures of anxiety at three points in time, meet this assumption because these Likert-type scales are at the ordinal level of measurement.

2. The data from a randomly selected sample are either (a) multiple observations from a single sample across more than two time periods or conditions or (b) blocks of matched subjects in which the subjects from a given block are each randomly assigned to one of the three or more conditions. This assumption implies that the researcher either has collected repeated measures on a single sample for more than two time periods or has matched sets of subjects who have been randomly assigned to three or more given conditions. The anxiety data consist of multiple observations across three time periods from a single sample of 10 children who took part in the intervention. These data, although not from a randomly selected sample, do meet the assumption of paired observations.

3. The subjects or blocks of subjects are independent; that is, the results within one block do not have an influence on the results within the other blocks. This third assumption means that no subject appears more than once and, therefore, does not appear within more than one block (or row). This would also imply that subjects who might have an undue influence on each other (e.g., husbands and wives, or twins) are not in separate blocks. Because our *blocks* (or rows) consist of single subjects and none of the subjects are related, our data meet this assumption.

Computer Commands

Figure 6.2 presents the SPSS for Windows commands used to generate the Friedman test. This dialogue box was opened by selecting the same items from the menu as for Cochran's *Q* test: *Statistics . . . Nonparametric Tests . . . k Related Samples.* The Friedman test was selected from the *Test Type* menu. To obtain descriptive statistics for the three time periods, the subcommand *Statistics* was highlighted and the items *Descriptive* (for

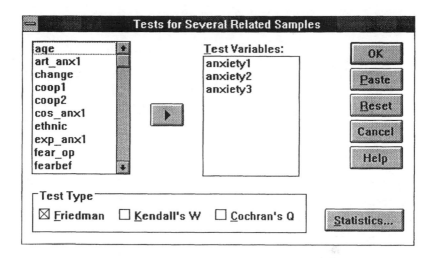

Figure 6.2. SPSS for Windows Computer Commands for the Friedman Test

means and standard deviations) and *Quartiles* (for the median) were selected.

Computer-Generated Output

Table 6.7 presents the syntax commands and computer-generated output that were obtained from SPSS for Windows. As with the previous tests, only the 10 children from the intervention group (Group = 0) were selected for this analysis (①). The syntax commands also indicate that the Friedman test has been requested (②) and that any missing values will be excluded from all requested analyses ("/MISSING LISTWISE"). In the generated output, the descriptive statistics, as requested, are presented first (③). These statistics suggest that there is a similar pattern for both the means and the medians (④) of the anxiety scores for the 10 children who took part in the intervention; that is, the children's average scores were highest at pretest, decreased following the intervention, and then increased somewhat just prior to administration of the preoperative medication.

When the Friedman test was undertaken, the resulting chi-square value was 8.15 with $df = 2$ (⑤). This is the same value as that which we obtained

Table 6.7 Syntax Commands and Computer-Generated Printout From SPSS for Windows for the Friedman Test

```
COMPUTE filter_$=(group = 0).   ①
VARIABLE LABEL filter_$ 'group = 0 (FILTER)'.
VALUE LABELS filter_$  0 'Not Selected' 1 'Selected'.

FORMAT filter_$ (f1.0).
FILTER BY filter_$.

EXECUTE .

NPAR TESTS
  /FRIEDMAN = anxiety1 anxiety2 anxiety3   ②
  /STATISTICS DESCRIPTIVES QUARTILES
  /MISSING LISTWISE.


              N       Mean      Std Dev  Minimum  Maximum ③
ANXIETY1     10     5.80000     1.39841    3.00     7.00
ANXIETY2     10     4.40000      .96609    3.00     6.00
ANXIETY3     10     4.90000      .87560    4.00     6.00


                                (Median) ④
                     25th        50th        75th
              N    Percentile  Percentile  Percentile
ANXIETY1     10     4.7500      6.0000      7.0000
ANXIETY2     10     3.7500      4.5000      5.0000
ANXIETY3     10     4.0000      5.0000      6.0000

- - - - - Friedman Two-Way Anova
     Mean Rank    Variable
  ⑦    2.70      ANXIETY1    preintervention anxiety
        1.45      ANXIETY2    post-test anxiety
        1.85      ANXIETY3    preoperative medication


        Cases         Chi-Square         D.F.     Significance
         10            8.1500 ⑤           2        1.70E-02 ⑥
```

when we hand-calculated the Friedman test. To determine whether this value is significant, it is necessary to compare the obtained significance level (⑥) to our prestated alpha level (e.g., .05). If the obtained significance is lower than our alpha, we will reject the null hypothesis of no differences among the medians for the three time periods. Because this is an overall test and not directional, we can examine the significance level directly without dividing it in half as we did for the directional Wilcoxon test in Chapter 5.

The resulting significance level was 1.70E-02 (⑥). This is a mathematical shorthand way of writing 1.70×10^{-2}. You will recall that $10^{-2} = .01$; therefore, our significance level is 1.70E-02 = 1.70×10^{-2} = $1.70 \times .01$ = .017. Our null hypothesis can be rejected because .017 is less than our stated level of alpha (.05). We can conclude, therefore, that at least one of the medians for the three time periods is significantly different from one of the others. To determine where the differences are located, it is necessary to undertake post hoc tests.

Post Hoc Comparisons to Determine the Locations of Differences

Two slightly different post hoc approaches that can be used to determine the differences among the time periods or conditions have been presented by Siegel and Castellan (1988) (see also Hettmansperger, 1984) and Munro and Page (1993). Because the two approaches generate different results in our data, they are examined individually here.

Comparing differences in average ranks. Once a determination has been made that the overall Friedman test is significant, post hoc tests can be undertaken that compare the differences in average ranks for all possible pairs (or certain conditions against a baseline) to determine where the differences lie (Hettmansperger, 1984; Siegel & Castellan, 1988). The null hypothesis of no differences in mean ranks of the pairs being examined (e.g., Time 1 vs. Time 2) will be rejected if the absolute value of these differences is greater than a specified critical value. That is, we would reject the null hypothesis if the following condition holds true:

$$|\overline{R}_1 - \overline{R}_2| \geq Z_{\alpha/[k(k-1)]} \sqrt{k(k+1)/(6N)}$$

where

R_1 = the mean rank for Time 1

R_2 = the mean rank for Time 2

$Z_{\alpha/[k(k-1)]}$ = the critical z value for $\alpha' = \alpha/[k(k-1)]$

k = the number of time periods or conditions

N = the number of cases.

In our hypothetical intervention example, the average ranks for Times 1-3 are 2.70, 1.45, and 1.85, respectively (Table 6.7, ⑦). Because $\alpha = .05$

and $k = 3$, the critical value for the z statistic is a z in which $\alpha' = .05/3(2) = .0083$. An examination of a table that presents the probabilities associated with the standard normal distribution (e.g. Siegel & Castellan, 1988, Table A) indicates that, for our study, $z = 2.39$. Our critical value for $|\overline{R}_1 - \overline{R}_2|$, therefore, is:

$$|\overline{R}_1 - \overline{R}_2| \geq 2.39\sqrt{3(4)/[6(10)]} \geq 2.39(.45) \geq 1.08$$

The absolute values for our three comparisons are as follows:

$$|\overline{R}_1 - \overline{R}_2| = |2.70 - 1.45| = 1.25$$

$$|\overline{R}_1 - \overline{R}_3| = |2.70 - 1.85| = 0.85$$

$$|\overline{R}_2 - \overline{R}_3| = |1.45 - 1.85| = 0.40.$$

The Time 1-Time 2 comparison (1.25) is the only one that is greater than our critical value of 1.08; therefore, we can conclude that, according to this post hoc approach, the intervention was effective in reducing the children's anxiety from preintervention to postintervention but that the significant reduction in anxiety was not maintained at the time of the administration of the preoperative medication.

Using Wilcoxon tests for post hoc comparisons. It is also possible to undertake nondirectional post hoc analyses for the Friedman test using Wilcoxon tests (Munro & Page, 1993). The critical alpha value is first adjusted using Bonferroni's inequality (i.e., adjusted $\alpha' = \alpha/k$ where $k =$ the number of tests undertaken and $\alpha =$ the original alpha level) to take into account the potential for increased Type I error. Note that, in this approach, α is divided by k, not $k(k-1)$, as in the first example. In our case, we would set our new alpha at .017 (.05/3 = .017) and then perform Wilcoxon tests on the three comparisons. We would reject the null hypothesis if the generated two-tailed significance level is less than .017.

Table 6.8 presents the z scores and p values for the three Wilcoxon comparisons. Unlike the results from the previous post hoc analyses, it is apparent that, with the Wilcoxon tests, *none* of our nondirectional pairwise comparisons is significant at $\alpha = .017$. The significance levels for all three comparisons are greater than .017. Even if we had selected a one-tailed test (assuming that we predicted the direction of the change in anxiety levels

Table 6.8 Post Hoc Analyses of the Time 1-Time 3 Data Using the
Wilcoxon Test

```
- - - - - Wilcoxon Matched-Pairs Signed-Ranks Test

   ANXIETY1   preintervention anxiety
with ANXIETY2   post-test anxiety

   Mean Rank      Cases

       4.75         8    - Ranks  (ANXIETY2 LT ANXIETY1)
  ④    7.00         1    + Ranks  (ANXIETY2 GT ANXIETY1)
                    1      Ties   (ANXIETY2 EQ ANXIETY1)
                   --
                   10      Total

        Z =   -1.8363            2-Tailed P =  .0663  ②
- - - - - Wilcoxon Matched-Pairs Signed-Ranks Test

   ANXIETY1   preintervention anxiety
with ANXIETY3   preoperative medication

   Mean Rank      Cases

       5.13         8    - Ranks  (ANXIETY3 LT ANXIETY1)
       4.00         1    + Ranks  (ANXIETY3 GT ANXIETY1)
                    1      Ties   (ANXIETY3 EQ ANXIETY1)
                   --
                   10      Total

        Z =   -2.1917            2-Tailed P =  .0284  ①
- - - - - Wilcoxon Matched-Pairs Signed-Ranks Test

   ANXIETY2   post-test anxiety
with ANXIETY3   preoperative medication

   Mean Rank      Cases

       5.00         1    - Ranks  (ANXIETY3 LT ANXIETY2)
       3.20         5    + Ranks  (ANXIETY3 GT ANXIETY2)
                    4      Ties   (ANXIETY3 EQ ANXIETY2)
                   --
                   10      Total

        Z =   -1.1531            2-Tailed P =  .2489  ③
```

prior to our post hoc analyses), only the Time 1 versus Time 3 comparison
(Table 6.8, ①) would have been significant ($.0284/2 = .0142 < .017$). The
Time 1 versus Time 2 (②) and Time 2 versus Time 3 ③ comparisons were
not significant ($p > .017$); thus, we would have concluded that our intervention changed the anxiety levels only from preintervention (Time 1) to
preoperative medication (Time 3). This conclusion appears contrary to
what we reached in the previous post hoc analysis, where Time 1 versus
Time 2 was the only significant group difference found.

Why would two post hoc approaches using the same data produce such different results? One reason is that each approach adjusts the alpha level slightly differently to accommodate for increased Type I error as a result of multiple tests using the same data. The two approaches are also nondirectional tests, although Siegel and Castellan (1988) offer the critical values for one-tailed tests as well (Table A_{II}, p. 320).

A second possible explanation for this discrepancy is that the Wilcoxon approach considers each pair of time periods individually and ranks data within time periods. The mean positive and negative ranks are, therefore, strongly affected by outliers, especially those in small samples. Note that there was one child in this small sample of 10 children whose anxiety scores increased considerably from Time 1 to Time 2 (Table 6.8, ④). This child substantially influenced the Wilcoxon results.

In contrast, the Siegel and Castellan (and Hettmansperger) post hoc approach considers the mean ranks of all three time periods generated from the Friedman test. These results are obtained by ranking the data first by row (i.e., within subject or matched group) and then averages by column (within time period or condition). Outliers do not have such a strong impact on the results because, if there are three conditions, the possible rank for any single observation is 1, 2, or 3.

Which post hoc analysis should be used? Because the Wilcoxon post hoc analyses are affected by small sample sizes in which there are apparent outliers, the researcher may want to use the Siegel and Castellan approach when faced with these limitations. This is what we will do with this hypothetical example. The disadvantage to this approach is that there is no handy computer program to undertake the analyses. The researcher will need to rely on a good calculator. This may be a bit tiresome if there are many observational periods or conditions. In contrast, the Wilcoxon post hoc tests are easily obtained by computer (see Chapter 5) and may be better used with larger samples in which outliers are not apparent. The researcher may also want to consider a one-tailed test if the direction of differences can be predicted prior to the post hoc analyses.

Presentation of Results

Table 6.9 offers a tabular presentation of the results of the Friedman test using the Siegel and Castellan approach to post hoc analyses. These data could also be presented in the text as follows:

The results of the Friedman test indicate that there was a significant difference in the median anxiety levels of the 10 children who took part in the intervention

Table 6.9 Suggested Presentation of Friedman Test Results

| | | | | | Friedman | |
Anxiety Scores	N	M	SD	Md	χ^2	p
Preintervention*	10	5.8	1.4	6.0		
Postintervention	10	4.4	0.9	4.5	7.20	0.03
Preoperative medication	10	4.9	1.1	5.0		

*Significantly different from postintervention, $p < .016$.

over the three time periods ($p = .017$). Post hoc analyses (Siegel & Castellan, 1988) with adjustment of the two-tailed level to .017 to accommodate increased Type I error indicated that there were significant decreases in the children's reported anxiety from preintervention ($Md = 6.0$) to postintervention ($Md = 4.5$). No other significant pairwise differences between time periods were obtained.

Advantages and Limitations of the Friedman Test

The Friedman test is a very versatile technique that can be used both with randomized block designs and with multiple observations of a single sample. It is especially useful when the dependent data being analyzed are continuous but their distributions are skewed.

There are, however, several potential drawbacks to this test. As indicated with the Wilcoxon test (Chapter 5), although the Friedman test is used to assess medians, it really is based on an assessment of ranks. Therefore, it is possible to obtain differences in directions between medians and mean ranks. It is also conceivable that the medians do not change but that the Friedman test yields a significant result. The researcher needs to carefully examine both the medians and mean ranks to be certain that the results make intuitive sense.

Although it is often referred to as the "Friedman two-way ANOVA by ranks," the Friedman test is restricted to within-group comparisons only. It is not possible to use this test for between-group comparisons across multiple time periods or conditions. This is a major disadvantage in clinical research, because it is not possible to make experimental-control group comparisons. While it is possible to analyze each of the groups independently and compare their results, there does not appear to be a nonparametric test available in the more popular statistical packages that provide a repeated-measures group by time interaction analysis with independent groups.

Parametric and Nonparametric Alternatives to the Friedman Test

The parametric counterpart to the Friedman test is the within-subjects repeated-measures ANOVA. There has been some question about the relative efficiency of the Friedman test compared to the F test for the repeated-measures ANOVA. Recall from Chapter 3 that the *power efficiency* of a test refers to the increase in sample size that is necessary to make one test (e.g., the Friedman test) as powerful as its rival (e.g., the F test) given a constant alpha level and fixed sample size for the rival test. Siegel and Castellan (1988) indicate that, compared to the F test, the power efficiency of the Friedman test is 64% when $k = 2$, increases to 80% for $k = 5$, and increases further to 87% for $k = 10$. This means that the discrepancy in sample size requirements decreases as the number of time periods or conditions increases. Hettmansperger (1984) argues that the Friedman test does not have as high a power efficiency relative to the F test in ANOVA as the Wilcoxon signed ranks test does with the paired t test. Zimmerman and Zumbo (1993) conducted a computer simulation to compare the Friedman test, the Wilcoxon signed ranks test, repeated measures ANOVA, and repeated measures ANOVA on ranks. These authors conclude that an ANOVA based on ranks may have more potential than the Friedman test in its sensitivity.

A second nonparametric test to which the Friedman test is functionally related is Kendall's W (Conover, 1980; Siegel & Castellan, 1988), a test that will be examined in detail in Chapter 9. Conover (1980) indicates that Kendall's W is a simple modification of the Friedman test and can be used in the same situations for which the Friedman test is applicable. It has been used in research, however, primarily as an assessment of "agreement in ranking" rather than as a test of differences among medians.

Examples From Published Research

Bertoti, D. B. (1988). Effect of therapeutic horseback riding on posture in children with cerebral palsy. *Physical Therapy*, *68*, 1505-1512.

Blegen, M. A., Goode, C. J., Johnson, M., Maas, M. L., McCloskey, J. C., & Moorhead, S. A. (1992). Recognizing staff nurse job performance and achievements. *Research in Nursing and Health*, *15*, 56-66.

Crawford, J. D., & McIvor, G. P. (1985). Group psychotherapy: Benefits in multiple sclerosis. *Archives of Physical Medicine and Rehabilitation*, *66*, 810-813.

McCain, G. C. (1992). Facilitating inactive awake states in preterm infants: A study of three interventions. *Nursing Research*, *40*, 359-363.

Philip, P. A., Ayyangar, R., Vanderbilt, J., & Gaebler-Spira, D. J. (1994). Rehabilitation outcome in children after treatment of primary brain tumor. *Archives of Physical Medicine and Rehabilitation, 75,* 36-39.

Summary

In this chapter, we have examined two nonparametric tests that are useful for assessing differences among observations across more than two time periods or conditions: Cochran's Q test and the Friedman test. To determine which of these two tests would be most appropriate for a specific set of data, it would be useful to review briefly the characteristics of the two tests.

Cochran's Q test is a useful statistical technique that can be used when the researcher is interested in evaluating change in dichotomous data across more than two time periods or conditions. It also can be used to examine agreement among raters or to evaluate relative difficulty of items on a test.

When the outcome data that are being evaluated across multiple time periods or conditions have some continuity attached to their meanings (i.e., they are at least at the ordinal level of measurement), the Friedman test is the preferred nonparametric statistic. It is a very versatile and robust statistic that can be used both with multiple measurements on a single sample (e.g., pretest, posttest, and follow-up) and on randomized block designs in which matched subjects in a block are randomly assigned to one of k conditions.

Both Cochran's Q and the Friedman test are overall tests of significance. Post hoc tests are needed, therefore, to determine where the specific differences among the groups lie. A disadvantage to both of these tests is that they can accommodate only within-subjects (or blocks) measurements. They are not useful for evaluating between-group differences, although some accommodations can be made to analyze such data. In the next two chapters, we will examine tests for independent groups. Chapter 7 will present nonparametric statistics that can be used with two independent groups: the Fisher exact test, the chi-square test, the median test, and the Wilcoxon-Mann-Whitney test. In Chapter 8, we will examine tests that accommodate more than two independent groups: the chi-square test for k independent samples and the Kruskal-Wallis one-way analysis of variance by ranks.

7 Tests for Two Independent Samples

- **Fisher exact test**
- **Chi-square test of independence**
- **Mann-Whitney *U* test**

In Chapters 5 and 6, we examined nonparametric tests that could be used when the data to be analyzed are dependent; that is, the data either are generated from repeated observations of a single sample or are obtained from matched samples. In health care research, we are often interested in comparing the outcomes obtained among groups or samples that are independent of one another, such as intervention and control groups, males and females, or persons of differing marital status. Two groups are considered to be *independent* if membership in one group excludes the possibility of membership in the second group. For example, a remarried respondent cannot also be listed as a divorced respondent, or a member of an intervention group cannot be a member of the control group. Independent samples can be obtained in one of two ways: (1) they are randomly drawn from mutually exclusive populations, such as a population of males and a population of females; or (2) they are randomly assigned to only one of several possible conditions, such as an experimental or control group.

In this chapter, we will examine three nonparametric tests that are available when the independent or predictor variable is *dichotomous*; that is, this variable is made up of two mutually exclusive groups or *independent samples* and there are no repeated measures. Two of the tests, the Fisher exact test and the chi-square test of independence, are used when both the independent and dependent variables are at the nominal level of measurement. The third test, the Wilcoxon-Mann-Whitney *U* test, is used when the independent variable is nominal and the dependent variable is at least at the ordinal level of measurement.

It should be noted that all these tests are used when the data are collected during one time period. The only time these tests can be used for repeated measures is when a difference score (e.g., Time 1 – Time 2) is calculated and these obtained difference scores are compared between the groups.

The Fisher Exact Test

The Fisher exact test is used to analyze data for which both the independent and dependent variables are dichotomous. It is especially useful when sample sizes are so small that the chi-square test for independent samples is inappropriate (McNemar, 1969). For example, in SPSS for Windows, the Fisher exact test is the default for the chi-square test of independence when the expected value of a 2×2 contingency table for every cell is less than 5, or when the total sample size is less than 20. The main purpose of such a test is to examine whether two populations differ from each other in the proportion of subjects who fall into one of two classifications. For example, we might be interested in comparing males and females with regard to their success (+) or failure (–) in treatment.

An Appropriate Research Question
for the Fisher Exact Test

The Fisher exact test has been used extensively in clinical research. A quick review of the health care literature for the 5-year period 1989-1994, for example, indicated that more than 250 clinical studies had included a Fisher exact test in their statistical analyses. Both the sample sizes and the research purposes varied considerably. For example, in the medical field, Graff-Radford, Godersky, and Jones (1989) used this statistic to assess outcomes in their prospective study of 30 elderly patients who had shunt surgery for symptomatic hydrocephalus. Levenson, Mishra, Hamer, and Hastillo (1989) used the Fisher exact test to examine the impact of denial on medical outcomes in 48 patients being treated for unstable angina. Goetz, Squier, Wagener, and Muder (1994) used the Fisher exact test to determine the incidence of infections and risk factors for nosocomial infections in 22 HIV-infected patients, and Hendrickse, Kusmiesz, Shelton, and Nelson (1988) used the statistic to compare outcomes from 5 versus 10 days of therapy for acute otitis media in a double blind study of 175 patients. Other studies that have used the same statistic to examine health care outcomes include Crozier, Graziani, Ditunno, and Herbison (1991), Mange, Marino, Gregory, Herbison, and Ditunno (1992), and Smiley and Paradise (1991).

Table 7.1 Example of Null and Alternative Hypotheses Suitable for the Fisher Exact Test and the Chi-Square Test for Two Independent Samples

---

Null Hypothesis

H_0: The proportion of children in the anxiety reduction intervention group who express fear of their upcoming operation immediately following the intervention will be no different from that of the control group; that is, the two variables, group membership and post-intervention fear, are independent.

Alternative Hypothesis

H_a: A smaller proportion of children who take part in the anxiety reduction intervention program will express fear of the operation immediately following the intervention than that of children who do not take part in the intervention; that is, the two variables, group membership and postintervention fear, are dependent.

Suppose, in our hypothetical intervention study, that we were interested in comparing the children in our intervention and control groups with regard to their reports of preoperative fear (yes or no) following the planned intervention. An example of a research question that could be answered using the Fisher exact test would be as follows:

Is there an association between group membership (i.e., intervention, control) and the children's expressed postintervention fear (yes, no) of their impending operation?

Null and Alternative Hypotheses

Table 7.1 illustrates the null and alternative hypotheses that would be suitable for use with a Fisher exact test given the research question outlined above. Note that the alternative hypothesis is directional. This would make sense, because we are predicting that there will be differences in proportions between the intervention and control groups in favor of the group that received the intervention. Because the alternative hypothesis is directional, we will be undertaking a one-tailed Fisher exact test.

Overview of the Procedure

To undertake the Fisher exact test, the dichotomous data to be analyzed are first cast into a 2 × 2 contingency table similar to Table 7.2, in which entries represent frequencies, not scores. Next, the probability is calculated

Table 7.2 Format for a 2 × 2 Contingency Table for the Fisher Exact Test

	Dependent Variable		
Independent Variable	*0*	*1*	*Total*
0	a	b	a + b
1	c	d	c + d
Total	a + c	b + d	N

of obtaining an arrangement of numbers in the four cells exactly like, or more disproportionate than, what was obtained in the original 2 × 2 table given similar row and column totals. For small sample sizes (e.g., $N \leq 15$), the hypergeometric distribution is employed, using the following formula (Hays, 1994; Siegel & Castellan, 1988):

$$p = \frac{(a+b)!(c+d)!(a+c)!(b+d)!}{N!a!b!c!d!}$$

where

p – the probability of obtaining the exact arrangement of variables presented in Table 7.2.

The Fisher exact test can become quite tedious to calculate by hand because p values need to be calculated not only for the exact frequencies but also for frequencies *more extreme* than what was obtained given similar row and column totals. Luckily, a number of textbooks (e.g., Daniel, 1990; Siegel & Castellan, 1988) present Fisher exact probability tables for sample sizes less than 15. The total generated p value is then compared to the preset alpha level (e.g., $\alpha = .05$). If this generated p value is less than the alpha, the null hypothesis that the generated arrangement of cell frequencies occurred purely by chance is rejected.

When the sample size is sufficiently large, a normal approximation to the hypergeometric distribution can be used (Daniel, 1990):

$$z = \frac{(a/[a+b]) - (c/[c+d])}{\sqrt{\bar{p}(1-\bar{p})}\,[(1/[a+b]) + (1/[c+d])]}$$

where

$$p = (a + c)/N$$

Daniel (1990) suggests that this normal approximation is appropriate when the frequencies in all four cells are ≥ 5. The null hypothesis of no association is rejected if the generated value of this z statistic is greater than the critical value of z at the prestated one- or two-tailed alpha (e.g., ± 1.96 for a two-tailed α = .05) or, similarly, if the generated p value for the z statistic is less than the prestated alpha.

In our hypothetical intervention study, we obtained the computer-generated contingency table presented in Table 7.2. Following our guidelines from Table 7.3, we would obtain the following values for a, b, c, and d: $a = 7$, $b = 3$, $c = 2$, and $d = 8$. This would make $\overline{p}$ = .45 since $\overline{p}$ = (a + c) / N = (7 + 2) / 20 = .45. Our z statistic, therefore, would be as follows:

$$z = \frac{(a/[a+b])-(c/[c+d])}{\sqrt{\overline{p}\,(1-\overline{p})\,[(1/[a+b]) + (1/[c+d])]}}$$

$$= \frac{(7/[7+3])-(2/[2+8])}{\sqrt{.45(1 - .45)\,[(1/[7+3]) + (1/[2+8])]}}$$

$$= \frac{.7 - .2}{\sqrt{.0495}} = 2.2473$$

This calculated z statistic, 2.25, is greater than the critical value of our one-tailed z statistic, 1.64, at α = .05. Similarly, a table of values for the z statistic indicates that the generated p value for this z is .012, which is less than α = .05. The null hypothesis of no association therefore will be rejected. The meaning of this conclusion will be discussed when we examine the computer printout.

An alternative to the z statistic for large sample sizes is the chi-square test for two independent samples, which will be reviewed later in this chapter. This statistic is the default in SPSS for Windows when the sample size is sufficiently large (i.e., when all expected cell values in a 2 × 2 table are greater than 5).

The Fisher exact test is usually considered to be a one-tailed test because it examines results that are as extreme as or more extreme than the obtained values in a specific direction (Hays, 1994; Siegel & Castellan, 1988). Hays (1994) suggests that when all row and column totals are equal, the two-tailed significance level can be obtained by doubling the p value. When the marginal totals are not equal (e.g., $[a + b] \neq [c + d]$), a different procedure must be used to obtain a two-tailed significance level. SPSS for Windows does not restrict the Fisher exact test to a one-tailed test but presents both one- and two-tailed probability values, adjusting for unequal marginals when necessary.

Critical Assumptions of the Fisher Exact Test

1. Both independent and dependent variables are dichotomous, with two mutually exclusive categories. Both variables being analyzed need to be dichotomous, with two mutually exclusive categories. In our hypothetical intervention study, both the grouping and fear variables are dichotomous, with two independent categories. If these variables were not dichotomous, their levels could be collapsed to meet this assumption.

2. The data to be analyzed have been obtained from a random sample and represent frequencies, not scores. As with all nonparametric statistical tests, the assumption is that the data have been obtained from a randomly selected sample. For this test, it is assumed that the values in each cell of the 2×2 table represent numbers of subjects, not scores. Although our hypothetical data are presented as counts, not scores, ours is a sample of convenience; therefore, we have only partially met this assumption.

3. Because this is an exact test of probability, it is necessary to know the number of cases in the marginals before the data are analyzed. Some statisticians have argued that the Fisher exact test is appropriate only when the marginal totals for both the independent and dependent variables are known (Daniel, 1990). This requirement suggests that the study needs to be experimental in design because in the case the researcher has control over the types of subjects entered into the study. For example, a researcher might want to be certain that a similar proportion of randomly assigned intervention or control group subjects possess a particular characteristic of interest (e.g., high blood pressure). In this case, both sets of marginals for group membership (intervention, control) and blood pressure (high, normal) could be fixed prior to the study.

Even with an experimental design, however, this assumption is not easily met. In our hypothetical study, we would need to know prior to collecting our data not only the number of children assigned to the intervention and control groups but also the total numbers of children who answered "yes" or "no" to the question concerning fear. This achievement is hardly realistic, given that the fear variable is not being manipulated, we hope, by the researcher. Norusis (1995b) suggests that this assumption is very restrictive, particularly in descriptive studies, and that there is disagreement among statisticians about the advisability of using the Fisher exact test when this assumption is not met.

Computer Commands

Figure 7.1 presents the SPSS for Windows commands that were used to generate the Fisher exact test. This dialogue box was obtained by selecting the following items from the menu: *Statistics . . . Summarize . . . Crosstabs*. To obtain the Fisher exact test, the *Chi-Square* box was selected from the *Statistics* menu in this dialogue box. Observed and expected values as well as row and column percentages also were requested under the *Cells* subcommand in the *Crosstabs* dialogue box.

Computer-Generated Output

Table 7.3 presents the syntax commands and computer-generated printout for the Fisher exact test that was generated from SPSS for Windows. The syntax commands (①) indicate that the command *Crosstabs* has been used to generate the Fisher exact test through its *Statistics = Chisq* subcommand (②). The generated table is set up such that, as requested in the syntax commands (③), the following values will be presented in the table: the count, expected value, row percentage, and column percentage. The first row of information in the table (③) represents the numbers of children in each cell ("Count"). The second row of values ("Exp Val") indicates the numbers of children that would be expected in the particular cell if the two variables, group membership and postintervention fear, were completely independent. The third ("Row Pct") and fourth ("Col Pct") rows represent the row and column percentages expressed as a percentage of all cases in that particular row or column.

From Table 7.3, it is possible to determine that, of the 11 children who expressed postintervention fear, 3 children (27.3%) were from the inter-

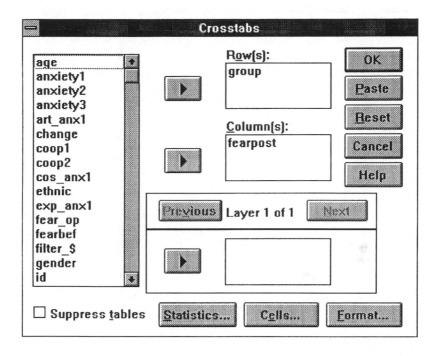

Figure 7.1. SPSS for Windows Commands for the Fisher Exact Test

vention group compared to 8 (72.7%) from the control group (④). In contrast, the expected values for each of these two cells are 5.5 (⑤), suggesting that the two variables, group membership and postintervention fear, are not independent. Looking at the third row of values, it is apparent that 30% (3 of 10) of the children in the intervention group expressed fear of their imminent operations, compared with 80% (8 of 10) of the children in the control group (⑥).

To determine whether the differences between the groups are significant, we need to look at the information presented for the Fisher's exact test. Because the one-tailed generated p value for this statistic, .03489, is less than our stated alpha (.05) (⑦), the null hypothesis of independence of the two variables, group membership and presence of postintervention fear, is rejected. The conclusion to be drawn is that the children in the intervention

Table 7.3 Syntax Commands and Computer-Generated Printout for the
Fisher Exact Test

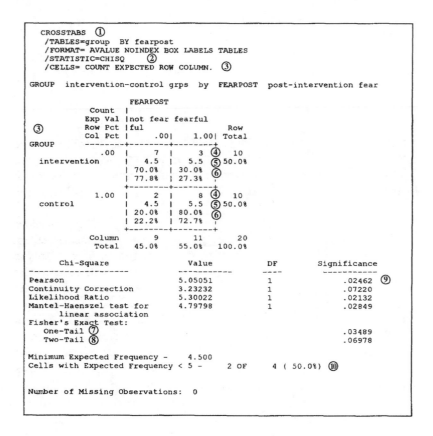

```
CROSSTABS   ①
 /TABLES=group  BY fearpost
 /FORMAT= AVALUE NOINDEX BOX LABELS TABLES
 /STATISTIC=CHISQ    ②
 /CELLS= COUNT EXPECTED ROW COLUMN.  ③

GROUP   intervention-control grps by  FEARPOST  post-intervention fear

                         FEARPOST
                 Count  |
                 Exp Val |not fear fearful
   ③           Row Pct |ful
                 Col Pct |    .00|    1.00|   Row
GROUP          --------+--------+--------+  Total
                 .00  |    7  |    3  ④    10
 intervention         |   4.5 |   5.5 ⑤ 50.0%
                      | 70.0% | 30.0% ⑥
                      | 77.8% | 27.3%
                 +--------+--------+
                1.00  |    2  |    8  ④    10
 control              |   4.5 |   5.5 ⑤ 50.0%
                      | 20.0% | 80.0% ⑥
                      | 22.2% | 72.7%
                 +--------+--------+
              Column      9      11      20
              Total    45.0%   55.0%  100.0%

        Chi-Square               Value       DF     Significance
  --------------------       ----------     ----    ------------
Pearson                        5.05051        1        .02462  ⑨
Continuity Correction          3.23232        1        .07220
Likelihood Ratio               5.30022        1        .02132
Mantel-Haenszel test for       4.79798        1        .02849
   linear association
Fisher's Exact Test:
   One-Tail ⑦                                          .03489
   Two-Tail ⑧                                          .06978

Minimum Expected Frequency -    4.500
Cells with Expected Frequency < 5 -     2 OF    4 ( 50.0%) ⑩


Number of Missing Observations:  0
```

group were less likely to express postintervention fear than the children in
the control group.

The two-tailed significance for the Fisher exact test is also presented
(⑧). Because there were equal numbers of children in the intervention and

Table 7.4 Suggested Presentation of Fisher Exact Test Results

	Postintervention Fear						
	Yes		No		Total		
Group Membership	N	%	N	%	N	%	p*
Intervention	3	27.3	7	77.8	10	50.0	
Control	8	72.7	2	22.2	10	50.0	.03
Total	11	55.0	10	45.0	20	100.0	

*The p value is for a one-tailed Fisher exact test.

control groups ($n = 10$), this value is, as Hays (1994) suggests, double the one-tailed p value. Had the numbers of children in each group been unequal (e.g., 10 children in the intervention and only 9 children in the control group), the two-tailed significance value would not have been twice the one-tailed value.

Note, too, that the computer-generated one-tailed p value, .03, is greater than the one-tailed p value (.012) for the z statistic for the Fisher exact test that had been hand-calculated (see above). These differences may be due to the small sample size and the resulting poor approximation to the hypergeometric distribution. Coincidentally, the p value for the z statistic (.012) is exactly half that of the two-tailed chi-square statistic (.02462) (⑨).

SPSS for Windows offers additional tests for contingency tables when the chi-square tests are requested. These tests include the Pearson chi-square, the chi-square test with a Yates correction, the likelihood ratio, and the Mantel-Haenszel test for linear association. Interpretation of these tests will be presented later in this chapter and in Chapter 8.

Presentation of Results

Table 7.4 is a suggested presentation of the results obtained from the Fisher exact test. The findings presented in this table could be interpreted and addressed in the text using the following statement:

The results of the Fisher exact test indicate that significantly fewer of the 10 intervention group children ($n = 3$) expressed postintervention fear than did the 10 control group children ($n = 8$) ($p = .03$).

Advantages and Limitations of the Fisher Exact Test

The Fisher exact test is an extremely useful and powerful statistic if the independent and dependent variables are at the nominal level of measurement. It is especially useful if either the sample size or the expected value of a particular cell is small (i.e., $N \leq 20$ or expected values < 5). If the sample size is larger than 20, calculation of the test by hand quickly becomes unwieldy and tedious. For that reason, Siegel and Castellan (1988) suggest that when the sample size is even greater than 15, the approximation of the chi-square test should be used. The Fisher exact test cannot be used when the variables being considered are continuous or are categorical with more than two levels.

A number of critics of the Fisher exact test have indicated that the test is too conservative (Hirji, Tan, & Elashoff, 1991; Overall & Hornick, 1982), that is, that it is less likely to correctly reject the null hypothesis. In response to this criticism, both Hirji et al. (1991) and Overall and Hornick (1982) present modified versions of the Fisher exact test that are purported to be more likely to correctly reject the null hypothesis than the unmodified version.

Parametric and Nonparametric Alternatives to the Fisher Exact Test

Because the Fisher exact test deals with 2×2 tables in which the variables being examined are at the nominal level of measurement, there is no parametric equivalent to this test. The chi-square test for two independent samples is a nonparametric alternative to the Fisher exact test when the sample size is sufficiently large and when the categorical variables have more than two levels. In addition, Berry and Mielke (1987) present an approach to using the Fisher exact test with 3×2 cross-classification tables and, as indicated, there have been suggested modified versions of the Fisher Exact test (Hirji et al., 1991; Overall & Hornick, 1982) that appear to be less conservative than the unmodified version.

Examples From Published Research

Crozier, K. S., Graziani, V., Ditunno, J. F., & Herbison, G. J. (1991). Spinal cord injury: Prognosis for ambulation based on sensory examination in patients who are initially motor complete. *Archives of Physical Medicine and Rehabilitation*, *72*, 119-121.

Goetz, A. M., Squier, C., Wagener, M. M., & Muder, R. R. (1994). Nosocomial infections in the human immunodeficiency virus-infected patient: A two year survey. *American Journal of Infection Control, 22*, 334-339.

Graff-Radford, N. R., Godersky, J. C., & Jones, M. P. (1989). Variables predicting surgical outcome in symptomatic hydrocephalus in the elderly. *Neurology, 39*, 1601-1604.

Hendrickse, W. A., Kusmiesz, H., Shelton, S., & Nelson, J. D. (1988). Five vs. ten days of therapy for acute otitis media. *Journal of Pediatric Infectious Diseases, 7*, 14-23.

Levenson, J. L., Mishra, A., Hamer, R. M., & Hastillo, A. (1989). Denial and medical outcome in unstable angina. *Psychosomatic Medicine, 51*(1), 27-35.

Mange, K. C., Marino, R. J., Gregory, P. C., Herbison, G. J., & Ditunno, J. F. (1992). Course of motor recovery in the zone of partial preservation in spinal cord injury. *Archives of Physical Medicine and Rehabilitation, 73*, 437-441.

Smiley, B. A., & Paradise, N. F. (1991). Does the duration of N_2O administration affect postoperative nausea and vomiting? *Nurse Anesthetist, 2*, 13-18.

The Chi-Square Test for
Two Independent Samples

The chi-square test for two independent samples (χ^2) is one of the most commonly used nonparametric statistics in health care research (Brown & Hayden, 1985; Pett & Sehy, 1996). Like the Fisher exact test, this test can be used to assess 2 × 2 contingency tables when both the independent and dependent variables are at the nominal level of measurement. Whereas the Fisher exact test is used for small samples, the chi-square test should be used when the total sample size is greater than 20 and the expected values for each cell are greater than 5. Unlike the Fisher exact test, the chi-square test for two independent samples can be used when the dependent variable has more than two categories. In Chapter 8, we will examine a chi-square test of association in which both independent and dependent variables are categorical with more than two levels.

An Appropriate Research Question for the
Chi-Square Two-Sample Test

The chi-square test for two independent samples has been used in numerous clinical research projects. For example, Von Roenn and colleagues (1988) used the chi-square test for two independent samples in their evaluation of sequential hormone therapy for 60 women with advanced breast cancer. Mann and colleagues (1986) used a similar test to determine the seroprevalence to HIV among 368 children ages 2 to 14 who were admitted to a hospital pediatric service in Zaire, and Pittinger, Maronian, Poulter, and Peacock (1994) examined the importance of margin status in breast-conserving therapy in 211 patients with Stage I or II breast cancer.

In our hypothetical study with only 20 children, the chi-square two-sample test would not be appropriate because two of the four cells of the 2×2 table have cells with expected frequencies of less than 5 (Table 7.3, ⑩). This is a violation of one of the assumptions of this test. To use the chi-square test, it would be necessary to increase the sample size. For the purposes of examining how the chi-square test might be used with these data, therefore, the number of children in our hypothetical intervention study has been magically increased from 20 to 30. A research question that could be answered using the chi-square test for two independent samples is similar to that for the Fisher Exact test:

What is the association between group membership (i.e., intervention vs. control) and the children's expressed fear of their impending operation (yes, no)?

Null and Alternative Hypotheses

The wording of the null and alternative hypotheses for the chi-square test for two independent samples is similar to that of the Fisher exact test (Table 7.1). The null hypothesis would state that the two variables, group membership and postintervention fear, are independent, whereas the alternative hypothesis would state that the two variables are dependent.

Overview of the Procedure

To undertake a chi-square test for two independent samples, the frequency data are first cast into a $2 \times k$ table (where k = the number of levels of the dependent variable). In our example, we would generate a 2×2 table because both variables, group membership and children's fear, are dichotomous. Next, the frequency of cases that fall within a particular cell is compared to the frequency that would be expected by chance if the two variables were independent. The expected number of cases in a particular cell (E_{ij}) is the product of the margin totals for a particular row (R_i) and column (C_j) divided by the total sample size (N):

$$E_{ij} = (R_i)(C_j)/N$$

From these data, a chi-square statistic is generated using the following formula:

$$\chi^2 = \sum_{i=1}^{r} \sum_{j=1}^{c} \frac{(O_{ij} - E_{ij})^2}{E_{ij}}$$

where

O_{ij} = the observed number of cases for a particular cell located in the ith row and jth column

E_{ij} = the expected number of cases for the same cell if the variables were independent

$\Sigma\Sigma$ = an indication that the fraction is summed across all rows (r) and columns (c).

If the data meet the assumptions of the chi-square test for two independent samples, this statistic is asymptotically distributed as a χ^2 with $(r-1)(c-1)$ degrees of freedom. If the resulting χ^2 statistic is sufficiently large, the null hypothesis of independence of variables is rejected. This χ^2 statistic, however, is distributed approximately as a χ^2 *only* if the expected frequencies are sufficiently large. A general rule of thumb is that if the degrees of freedom are equal to 1 (i.e., it is a 2×2 contingency table), all expected frequencies for the table should be ≥ 5. For larger tables ($df > 1$), no more than 20% of the cells should have expected frequencies of less than 5.

Critical Assumptions of the Chi-Square Test for Two Independent Samples

1. The data being analyzed must be frequency data, not scores. Like the Fisher exact test, the data being examined must consist of counts, not scores. Our data, group membership and children's fear, meet this assumption.

2. The variables being examined are categorical, with mutually exclusive levels. For the chi-square test for two independent samples, the independent variable must be dichotomous. The dependent variable, however, may have more than two levels. Each observation must be assigned to one and only one cell. This means that the cells are mutually exclusive and no subject has contributed to more than one cell. Because both the group membership (intervention, control) and fear (yes, no) variables are dichotomous, with mutually exclusive levels, this assumption is met.

3. The observations must be independent of one another. This assumption implies that no pair of observations can have any influence on another pair of observations. This is not a test for repeated observations of the same subject. Because, in our hypothetical study, each subject has only one pair of values, group membership and expressed fear, this assumption has been met.

The assumption of independence is very important because if the observations are dependent in any way, they can have dramatic effects on the researcher's decision regarding the null hypothesis. Violation of the assumption of independence could occur when a subject is asked to provide multiple responses to a particular question. For example, in our hypothetical study, a mother might be requested to list all the prescription drugs her child is taking, or a record might be kept of the types of side effects the child is experiencing. In both instances, although the types of drugs taken and side effects could be listed in a contingency table, the cells are no longer independent because the respondent's information could appear multiple times in the table. A chi-square test of independence therefore would be inappropriate.

4. If the two variables being examined are dichotomous, resulting in a 2×2 contingency table with df = 1, all expected frequencies for the table should be 5. For larger tables where df > 1, no more than 20% of the cells should have expected frequencies of less than 5. This rule is important because, as indicated earlier, the chi-square statistic is distributed approximately as a χ^2 only if the expected frequencies are sufficiently large. We will examine the extent to which the data from our hypothetical study meet this rule when we examine the computer printout.

Computer Commands

As for the Fisher exact test, the *Statistics . . . Summary . . . Crosstabs* commands in the SPSS for Windows menu generate the chi-square test for two independent samples (Figure 7.1). Once the *Crosstabs* dialogue box has been selected, open the *Statistics* menu and click on the boxes for the chi-square test and measures of association for 2 × 2 tables, such as the contingency and phi coefficients. Other useful information can be obtained by opening the *Cells* menu and requesting observed and expected frequencies, row and column percentages, and standardized residuals.

Computer-Generated Output

Table 7.5 presents the syntax commands and computer-generated printout for the chi-square test for two independent samples obtained from the SPSS for Windows *Crosstabs* commands. The syntax commands (①) outlines all that we have requested. In the printout, we are presented with the actual frequency ("Count"), expected frequency ("Exp Val"), percentages expressed in terms of row totals ("Row") and column totals ("Column"), and the standardized residuals ("Std Res") (②). As we will see later in this discussion, the values of the standardized residuals are especially helpful in determining which cells are influencing the outcomes observed.

Table 7.5 indicates that 10 children in the intervention group expressed no fear concerning their upcoming operation compared to only 4 children in the control group (③). If the two variables were independent, the expected number of children in each nonfearful group would be 7 (E_{ij} = $[R_i][C_j]/N$ = $[15][14]/30$ = 7). The expected number of children in the fearful groups would be 8 (E_{ij} = $[16][15]/30$ = 8) (④). The small discrepancy (+1.1 and −1.1) between the actual counts and expected values for each of the four cells (⑤) suggests that the two variables, group membership and postintervention fear, may not be independent.

To determine whether the differences between the observed and expected values are sufficiently large to reject the null hypothesis of independence of variables, we need to examine the generated Pearson chi-square and its significance level (⑥). If the significance level (*p* value) is less than our prestated alpha level, we will reject the null hypothesis. In our case, the χ^2 value is 4.82. This was obtained by using the formula outlined above:

$$\chi^2 = \sum_{i=1}^{r} \sum_{j=1}^{c} \frac{(O_{ij}-E_{ij})^2}{E_{ij}} = \frac{(10-7)^2}{7} + \frac{(5-8)^2}{8} + \frac{(4-7)^2}{7} + \frac{(11-8)^2}{8} = 4.82$$

Table 7.5 Computer-Generated Printout From SPSS for Windows for the Chi-Square Test for Two Independent Samples

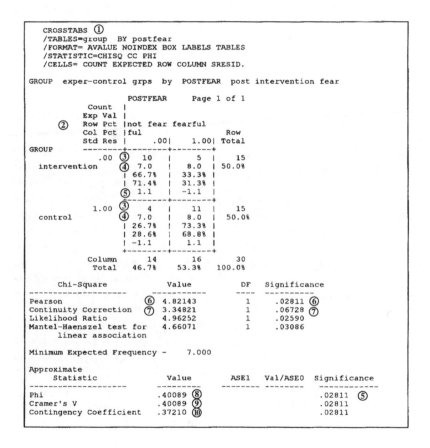

Because the generated p value (.02811, ⑥) for this χ^2 is less than our prestated alpha ($\alpha = .05$), we will reject the null hypothesis. The conclusion to be drawn is that the two variables, group membership and postintervention fear, are significantly associated or dependent.

Hinkle, Wiersma, and Jurs (1994) point out that the chi-square distribution (χ^2) is a continuous distribution with no breaks in its continuity. A serious problem occurs, therefore, when the expected frequencies in any of the cells for a 2×2 table are less than 5 because the resulting sampling distribution for this χ^2 could depart substantially from continuity. To improve the approximation to a χ^2 distribution when categorical data are being analyzed, an adjustment to the χ^2 has been suggested: the Yates correction for continuity (Cochran, 1954). The Yates correction factor for continuity is as follows:

$$\chi^2 = \sum \sum \frac{(O_{ij} - E_{ij} - .5)^2}{E_{ij}}.$$

Although the Yates correction has the advantage of making adjustments for the fact that there are categorical data, there are distinct disadvantages to its use. Hinkle and associates (1994) warn that, for a 2 × 2 table, the generated χ^2 value with the Yates correction is extremely conservative. This results in a substantial loss of power and a tendency to incorrectly fail to reject the null hypothesis. For that reason, these authors recommend ignoring the continuity correction for 2 × 2 tables. Norusis (1995b) supports this stance. Siegel and Castellan (1988) recommend using the Yates correction when the sample size is ≥ 40, arguing that the correction considerably improves the approximation to the χ^2 distribution.

In Table 7.5, we can see that the problem raised by Hinkle and colleagues (1994) has occurred in our hypothetical study. The Yates continuity correction (⊘) has resulted in a more conservative χ^2 value (3.348) that is no longer significant ($p = .06728$, ⊘) at our level of alpha. If we had used the χ^2 value that was corrected for continuity, therefore, we would have failed to reject the null hypothesis and would have, perhaps falsely, concluded that the two variables, group membership and postintervention fear, were independent (i.e., not associated).

Analyzing Residuals to Determine Which Cells Are Influencing the Outcome

The chi-square statistic is an overall test of independence between two variables. Given a significant result, this statistic does not indicate which cells appear to be major contributors to the observed differences. There are several approaches that can be used to determine which cells are influencing the outcomes observed. In Chapter 8, we will examine an approach suggested by Siegel and Castellan (1988) to analyze directions of influence in tables that are larger than 2 × 2. In this chapter, we will examine the residuals to determine which cells appear to be influencing the value of the χ^2 statistic obtained.

The difference between the actual and expected values represents the *residual* (i.e., what is "left over"). In SPSS for Windows, it is possible to request three types of residuals: unstandardized, standardized, and adjusted unstandardized. Hinkle and associates (1994) suggest that the *standardized* residuals be used to examine which cell(s) appear to be influencing the outcomes observed. To obtain a standardized residual for each category, a value similar to a z statistic is obtained. The difference between the observed and expected values of a cell is obtained and divided by the square root of the expected value:

$$\text{standardized residual} = \frac{O - E}{\sqrt{E}}$$

For Cell 1 in Table 7.5, the standardized residual would be $(10 - 7)/2.65 = 1.1$ (⑤). An absolute value > 2 of a standardized residual implies that the cell in question is an important contributor to the significant χ^2 value (Haberman, 1984; Hinkle et al., 1994).

In our hypothetical example, all the absolute values of the standardized residuals are similar (1.1), and no value exceeds 2 (Table 7.5, ⑤). We would conclude, therefore, that there is no one cell that is unduly influencing the generated χ^2 value.

Determining the Strength of Association Between Two Variables

The generated χ^2 statistic and its p value provide little information concerning the strength of the association between the two variables being examined. It is also not possible to compare χ^2 values across studies of different sample sizes because the size of the statistic is directly influenced by the size of the sample: An increase in sample size will most likely increase the value of the χ^2. It is possible, therefore, to obtain statistically significant results from the chi-square test for two independent samples that may or may not be of clinical interest to the researcher.

Several commonly used measures of association that are generated from the chi-square statistic provide the researcher with information concerning the strength of the association between two categorical variables (e.g., the phi coefficient, the contingency coefficient, and Cramér's *V*). One useful statistic for 2×2 tables is the *phi coefficient*. This statistic is examined in

Table 7.6 Suggested Rule of Thumb for Interpreting the Phi Coefficient for 2 × 2 Tables

Size of Phi Coefficient	r^2	Strength of Association
.90 to 1.00	.81 – 1.00	Very strong
.70 to .89	.49 – .80	Strong
.50 to .69	.25 – .48	Moderate
.30 to .49	.09 – .24	Low
.00 to .29	.00 – .08	Weak

greater detail in Chapter 9. In this chapter, the interpretation of the phi coefficient in view of a significant χ^2 statistic will be discussed briefly.

The Phi Coefficient

When a significant χ^2 for a 2 × 2 contingency table has been obtained, the phi coefficient can be used to determine the strength of association between the two variables. This coefficient is obtained by taking the square root of a significant χ^2 value which has been divided by the total sample size:

$$\phi = \sqrt{\frac{\chi^2}{N}}$$

For a 2 × 2 table, the values for the phi coefficient range between 0 and 1.00, with 0 indicating no relationship and 1.00 indicating a perfect relationship. If the categories of the independent and dependent variables in the 2 × 2 table have been coded 0 and 1, the phi coefficient is related to the Pearson product-moment correlation: $\phi = |r_{xy}|$ (Chapter 9). For interpretive purposes, therefore, the strength of relationship between two categorical variables could be viewed in a context similar to what Hinkle et al. (1994) suggest for a Pearson r (Table 7.6); that is, phi coefficients that range from 0 to .49 suggest a relationship that is of weak or low strength, .50 to .69 is a relationship of moderate strength, .70 to .89 indicates a strong relationship, and .90 to 1.00 suggests a very strong relationship.

The phi coefficient for our hypothetical study (Table 7.5, ⑧) was calculated as follows:

$$\phi = \sqrt{\frac{\chi^2}{N}} = \sqrt{\frac{4.82}{30}} = .40$$

According to the Hinkle et al. (1994) criteria (Table 7.6), the strength of association between group membership and postintervention fear is at best, low, a result that is not something on which a researcher might want to base a major funding application to the National Institutes of Health.

The Contingency Coefficient and Cramér's V

When one dimension of the table is greater than 2 (i.e., $2 \times k$), the value of the phi coefficient may not lie between 0 and 1 because the χ^2 value can be greater than the sample size. In that instance, the use of the *contingency coefficient (C)* has been suggested, where

$$C = \sqrt{\frac{\chi^2}{\chi^2 + N}}.$$

Although the value of this contingency coefficient will always lie between 0 and 1, it will never attain the upper limit of 1, and its maximum value depends on the number of rows and columns in the table (e.g., 0.87 for a 4×4 table) (Norusis, 1995b). For that reason, Cramér developed *Cramér's V*, a variant of the contingency coefficient:

$$V = \sqrt{\frac{\chi^2}{N(L-1)}}$$

where

N = the sample size

L = the smaller of the number of rows and columns.

This statistic will always range between 0 and 1 and, for a 2×2 table, will have the same value as the phi coefficient. In our hypothetical study (Table 7.5), the phi coefficient (⑧) and Cramér's V (⑨) are equivalent and have the same value. The contingency coefficient (⑩) is slightly smaller because of the larger denominator. Note that all coefficients ⑤ are significant at the same level as the χ^2 statistic ($p = .028$, ⑥).

Table 7.7 Suggested Presentation of Results for the Chi-Square Test for
Two Independent Samples

| | Postintervention Fear | | | | | | | |
| | Yes | | No | | Total | | χ^2 | p* |
Group Membership	N	%	N	%	N	%		
Intervention	5	31.3	10	71.4	15	50		
Control	11	68.8	4	28.6	15	50	4.82	.03
Total	16	53.3	14	46.7	30	100		

Presentation of Results

Table 7.7 suggests how the results obtained from the chi-square test for
two independent samples could be presented. These results could also be
interpreted and presented in the text as follows:

> The results of the chi-square test for two independent samples indicates that
> the 15 children in the intervention group expressed a lower rate of postinter-
> vention fear ($n = 5$) than did the 15 children in the control group ($n = 11$) (χ^2
> $= 4.82, p - .03$). The strength of the relationship between group membership
> and postintervention fear was low ($\phi = .40$), suggesting that other factors
> besides the clinical intervention may be influencing children's postinter-
> vention fear.

Advantages and Limitations of the Chi-Square Test
for Two Independent Samples

The chi-square test for two independent samples is a very popular test
that is easily understood and interpreted, especially when the data are cast
into a 2 × 2 table. When the contingency table is larger, it is more difficult
to determine which cells are most strongly influencing the significant χ^2.
In Chapter 8, we will examine ways to partition larger contingency tables
to determine where the major influences lie. Two additional disadvantages
to the chi-square test for two independent samples are that it is extremely
sensitive to departures from the assumption concerning expected frequencies, especially for 2 x 2 tables, and it is insensitive to order in the
categorical variables.

Parametric and Nonparametric Alternatives to the Chi-Square Two-Sample Test

Because the chi-square test for two independent samples is used when both independent and dependent variables are categorical, there is no parametric alternative to this test. Because the chi-square test is extremely sensitive to both sample size and expected frequencies, Cochran (1954) and Siegel and Castellan (1988) recommend that if the sample size is ≤ 20 and the data to be analyzed are in a 2 × 2 table, the Fisher exact test should be used instead of the chi-square test for independent samples. If the sample size is between 20 and 40, the chi-square test should be used for a 2 × 2 table if all the expected values in the cells are greater than 5. If there are cells that have expected frequencies less than 5, the Fisher exact test is recommended. When the sample is greater than 40, the Yates correction should be used for a 2 × 2 table (Cochran, 1954; Ludbrook & Dudley, 1994; Siegel & Castellan, 1988).

For 2 × k tables, both Cochran (1954) and Siegel and Castellan (1988) indicate that the chi-square test can be used if no more than 20% of the cells have expected frequencies of less than 5. If the data do not meet this assumption, it is possible to collapse cells so that a smaller table is generated. The choice of which cells should be collapsed together, however, requires thoughtful consideration, a discussion of which can be found in Chapter 8.

As indicated above, for 2 × 2 tables, the phi statistic generated from the χ^2 is related to the Pearson product-moment correlation provided that the cells for these categorical variables are coded as 0s and 1s. A correlation coefficient that is generated from these data can be interpreted as a Pearson product-moment correlation. For 2 × k tables, however, the chi-square test for two independent samples is not sensitive to order. For that reason, other nonparametric tests may be better alternatives when the dependent variable has some presumed order. In Chapter 8, we will examine the use of a Mantel-Haenszel chi-square test when the categorical data have some presumed order. In the final sections of this chapter, we will examine the use of the Wilcoxon-Mann-Whitney U test when the dependent variable is ordinal.

Examples From Published Research

Mann, J. M., Bila, K., Colebunders, R. L., Kalemba, K., Khonde, N., Bosenge, N., Nzilambi, N., Malonga, M., Jansegers, L., & Francis, H., et al. (1986). Natural history of human immunodeficiency virus infection in Zaire. *Lancet, 2*, 707-709.

Pittinger, T. P., Maronian, N. C., Poulter, C. A., & Peacock, J. L. (1994). Importance of margin status in outcome of breast conserving surgery for carcinoma. *Surgery,* *116,* 605-609.

Von Roenn, J. H., Bonomi, P. D., Gale, M., Anderson, K. M., Wolter, J. M., & Economou, S. G. (1988). Sequential hormone therapy for advanced breast cancer. *Seminars in Oncology, 15,* 38-43.

The Wilcoxon-Mann-Whitney U test

The Fisher exact test and the chi-square test for two independent samples are used when both the independent and dependent variables being examined are categorical. Because these two tests are not sensitive to order, they are not ideal measures to use when the dependent variable has some order to it. The next test that we will examine is the Wilcoxon-Mann-Whitney U test. This powerful nonparametric test is very useful when the dependent variable to be examined is continuous but either it does not meet the assumptions of a parametric test or the researcher wishes to generalize to more than just a "normally distributed" population.

This nonparametric test for two independent samples was independently developed by both Wilcoxon (1945) (the Wilcoxon W ranked sum statistic) and Mann and Whitney (1947) (the Mann-Whitney U statistic), thus the origins of its current name, the Wilcoxon-Mann-Whitney U test (Conover, 1980; Neter et al., 1993). This test is not to be confused with the Wilcoxon signed ranks test (Chapter 5), which is used with dependent samples. To avoid possible confusion, this test will be referred to in this chapter as the *Mann-Whitney test.*

The Mann-Whitney test is used when the researcher wants to determine whether two independent samples are from the same population. Like its parametric counterpart, the independent t test, the Mann-Whitney test compares measures of central tendency between two independent groups. Unlike the t test, the Mann-Whitney test uses medians for comparison, not means.

An Appropriate Research Question for the Mann-Whitney Test

The Mann-Whitney test is often used to compare two groups with regard to a given criterion that is measured on at least an ordinal scale, such as depression, wellness, or age. This test also can be used to compare change scores for two populations to determine whether these change scores come from the same population. That is, the researcher may want to compare the changes in a dependent variable (e.g., perceived quality of life) in two groups (e.g., intervention and control) across two periods of time. To do this, the researcher first creates a new variable (e.g., QOLDIFF, the difference between quality of life at Time 2 and Time 1) that represents a change score (e.g., QOLDIFF = QOL2 - QOL1). The Mann-Whitney test is then run on the new dependent variable, QOLDIFF, and the independent variable, which is dichotomous.

In our hypothetical example, we might be interested in comparing the intervention and control groups of children with regard to nurses' evaluations of the children's anxiety levels shortly following administration of the intervention to the control group. This variable, postintervention anxiety, has been measured on a continuous 7-point ordinal scale on which 1 = *not at all anxious* and 7 = *very anxious*. A possible research question that could be answered using the Mann-Whitney test is as follows:

Are the children who receive the anxiety reduction intervention (intervention group) perceived by the nurses to have lower postintervention anxiety than the children who do not receive the intervention (control group)?

Null and Alternative Hypotheses

The Mann-Whitney test is one of the most commonly used nonparametric tests in the health care research literature (Pett & Sehy, 1996). A cursory CD-ROM search of the Medline, CINAHL, and Psych Lit databases for the period 1990-1995, for example, yielded more than 850 publications— from a variety of disciplines and using a broad range of sample sizes—that reported using at least one Mann-Whitney test in their statistical analyses. These projects included a study of needle sharing and participation in a syringe exchange program among 131 injecting drug users (Hartgers, van Ameijden, van den Hoek, & Coutinho, 1992), the effects of developmental care on behavioral organization in 45 very low-birthweight infants (Becker, Grunwald, Moorman, & Stuhr, 1993), a study of the effects of rotation therapy on pulmonary complications in 69 liver transplant patients (Whiteman, Nachtmann, Kramer, Sereika, & Bierman, 1995), and an examination of improvements in quadriceps strength following an exercise program for 55 institutionalized elderly persons (McMurdo & Rennie, 1994).

In our hypothetical study of the impact of an anxiety reduction intervention program on children who have been hospitalized for minor surgery, we have indicated from our research question that we are interested in comparing the intervention and control groups with regard to their post-

Table 7.8 Example of Null and Alternative Hypotheses Suitable for Use
With the Mann-Whitney Test

Null Hypothesis

H_0: There will be no difference between the experimental and control groups of children
with regard to their median postintervention anxiety scores; that is, the two groups
come from the same population.

Alternative Hypothesis

H_a: The group of children who participate in the anxiety reduction intervention program
will have lower median levels of postintervention anxiety than the children in the
control group; that is, the groups come from different populations.

---

intervention anxiety scores. An example of null and alternative hypotheses
that could be answered using a Mann-Whitney test is given in Table 7.8.
Note that the alternative hypothesis is directional; therefore, our Mann-
Whitney test will be one-tailed.

We also could have compared the change in anxiety scores from preintervention to postintervention for the intervention and control groups. If
we chose that route, we would need to first create a new variable to
represent the change scores. In SPSS for Windows, this is accomplished
by clicking on the commands *Transform . . . Compute* and defining the new
variable (e.g., AnxChg = Anxiety1 – Anxiety2). The Mann-Whitney test is
then run on this new variable, AnxChg.

Overview of the Procedure

To calculate a Mann-Whitney test, the data for both groups are first
combined into a single sample, and their scores on the dependent variable
are ranked from lowest to highest without regard to their position on the
independent variable. Tied observations are assigned the average of the tied
ranks. Next, the independent groups are separated out, and the sum of the
assigned ranks for each group is calculated. If the null hypothesis that the
groups come from the same population is true, the sum of the ranks for both
groups should be similar. If the sum of the ranks for one group is very
different from that for the other, there is evidence to suspect that the two
samples did not come from the same population. The null hypothesis is
rejected if the sum of the ranks for one group is sufficiently larger than that
for the second group.

The statistic that is generated to determine whether the null hypothesis is to be rejected is the U statistic. This statistic takes into account both the measure of central tendency and the distribution of the scores and is defined as the smaller of U_1 and U_2, where:

$$U_1 = n_1 n_2 + \frac{n_1 (n_1 + 1)}{2} - R_1$$

$$U_2 = n_1 n_2 + \frac{n_2 (n_2 + 1)}{2} - R_2$$

where

n_1 = the number of observations in Group 1

n_2 = the number of observations in Group 2

R_1 = the sum of the ranks assigned to Group 1

R_2 = the sum of the ranks assigned to Group 2

The sampling distribution of this U statistic is known and, like the t distribution or the chi-square distribution, can be used for testing hypotheses. When the sample size is large, this U distribution quickly approaches a normal distribution with a mean, $\mu_v = n_1, n_2 / 2$, and standard deviation, $\sigma_u = \sqrt{n_1, n_2 (n_1 + n_2 + 1)/12}$. There is, however, some disagreement in the literature as to how large is "large". Neter and associates (1993) indicate that the approximation is adequate when both groups have sample sizes of 10 or more. Hinkle et al. (1994) argue that each group should have a sample size larger than 20.

The decision rule for the Mann-Whitney test is that, for small sample sizes, the null hypothesis is rejected if the obtained value of U (the smaller of U_1 or U_2) is *smaller* than the critical value of U at the predetermined level of alpha (e.g., $\alpha = .05$). For larger sample sizes, a z statistic is then generated (Hinkle et al., 1994; Neter et al., 1993; Siegel & Castellan, 1988):

$$z = \frac{U - \mu_u}{\sigma_u} = \frac{U - n_1 n_2 / 2}{\sqrt{n_1 n_2 (n_1 + n_2 + 1)/12}}$$

The null hypothesis is rejected if the absolute value of the obtained z value is *greater* than the absolute value of the critical z at the prespecified alpha level.

The reason for this apparent contradiction in rejection rules between the small and large sample statistics is that the U statistic for smaller sample sizes examines the *minimum* value of U_1 or U_2 and compares its minimum value to a critical minimum value that would occur if the null hypothesis were true. In contrast, for larger samples, the generated z statistic is compared to a critical z value at a prespecified alpha level.

Critical Assumptions of the Mann-Whitney Test

1. The independent variable is dichotomous, and the scale of measurement for the dependent variable is at least ordinal. The Mann-Whitney test is used when there are two groups and the variable being examined is at least at the ordinal level of measurement. The grouping variable for our hypothetical study has two levels (intervention, control), and our post-intervention anxiety variable is measured on a 7-point ordinal Likert-type scale. We have, therefore, met this assumption.

2. The data consist of a randomly selected sample of independent observations from two independent groups. It is assumed that the data have been randomly selected, that there are no repeated observations, and that the two levels of the independent variable are mutually exclusive. In our hypothetical study, the independent variable, group membership, consists of two mutually exclusive groups. The dependent variable, postintervention anxiety, consists of observations of the 30 children wherein no subject appears more than once in the data set and there are no repeated observations. Even though the children were randomly assigned to the intervention and control groups, however, the initial sample was one of convenience; therefore, the assumption of random selection has not been met.

3. The population distributions of the dependent variable for the two independent groups share a similar unspecified shape but with a possible difference in measures of central tendency. The Mann-Whitney test assumes that the two levels of the independent variable share a similar shape with regard to the distribution of the dependent variable, but this shape need *not* be bell-shaped or normal. In our hypothetical intervention study, we will need to examine the extent to which our data have similarity of

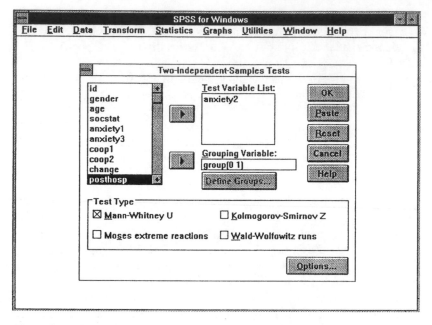

Figure 7.2. SPSS for Windows Commands for the Mann-Whitney Test

distribution shapes. We will check this assumption using the Kolmogorov-Smirnov (K-S) two-sample goodness-of-fit test (Chapter 4) when we undertake the computer analysis of the Mann-Whitney test.

Computer Commands

To obtain a Mann-Whitney test in SPSS for Windows, it is necessary to open the Nonparametric dialogue box for two independent samples (Figure 7.2). This is done by clicking on the following items in the menu: *Statistics . . . Nonparametric Tests . . . 2 Independent Samples.* The dependent variable is postintervention anxiety (*Anxiety2*) and the independent variable is our grouping variable (*Group*). The two levels of this categorical independent variable are defined by clicking on the *Define Groups* subcommand and indicating that all members of Group 1 (the intervention group) have been assigned the value of 0 and all members of Group 2 (the control group) have been assigned the value of 1. The *Mann-Whitney* test is selected from the *Test Type* menu.

Assessing the Shapes of the Distributions

Assumption 3 of the Mann-Whitney test states that the population distributions of the dependent variable for the two independent groups share a similar but unspecified shape. We saw in Chapter 4 that there are two ways to assess the extent to which our data meet this assumption: (1) examine visually the histograms of the dependent variable within the various levels of the independent variable, or (2) use the Kolmogorov-Smirnov two-sample test to check for similarity of distributions. The K-S two-sample test is quite easy to generate here because it uses the same SPSS for Windows dialogue box as the Mann-Whitney test. To ascertain the extent to which our data meet Assumption 3, we merely need to select the *Kolmogorov-Smirnov* test in addition to the *Mann-Whitney* from the *Test Type* group.

Computer-Generated Output for the Mann-Whitney Test

Table 7.9 presents the syntax commands and computer-generated output for both the K-S and the Mann-Whitney tests. The syntax commands, which can be used with earlier versions of SPSS, indicate that the Kolmogorov-Smirnov (K-S) (①) and Mann-Whitney U (M-W) (②) tests have been requested. The computer-generated printout for the Kolmogorov-Smirnov two-sample test informs us that the p value obtained (.400) (③) is larger than our prestated alpha ($\alpha = .05$); therefore, we cannot reject the null hypothesis, which, for the K-S two-sample test, states that the distributions for both groups are similar. The conclusion is, therefore, that we have met the assumption of similarity of distributions. To describe the characteristics of these distributions, it would be useful to examine the histograms of the two groups. These procedures, along with the K-S two-sample test, are described in detail in Chapter 4.

The computer-generated output for the Mann-Whitney test is also presented in Table 7.9. We are first presented with the average ranks for both groups (④). The *mean rank* is presented instead of the *sum of the ranks (R_1 and R_2)* to allow comparisons of ranks even if the groups are not equal in size. Note that the mean rank for the intervention group is smaller than the mean rank for the control group (④). Because this ordinal-level variable was scored such that 1 = *not at all anxious* and 7 = *very anxious*, the difference in these mean ranks suggests that the intervention group had lower postintervention anxiety scores than the control group.

Table 7.9 Syntax Commands and Computer-Generated Printout for the Mann-Whitney Test

```
NPAR TESTS
  /K-S= anxiety2    BY group(0 1)   ①
  /M-W= anxiety2    BY group(0 1)   ②
  /Missing Analysis.

- - - - - Kolmogorov - Smirnov 2-Sample Test

    ANXIETY2   post-test anxiety
  by GROUP     exper-control grps

       Cases

          10  GROUP =  .00  intervention
          10  GROUP = 1.00  control
          --
          20  Total

Warning - Due to small sample size, probability tables should be consulted.

              Most extreme differences
      Absolute         Positive         Negative      K-S Z    2-Tailed P
       .40000           .40000           .00000        .894      .400  ③

- - - - - Mann-Whitney U - Wilcoxon Rank Sum W Test

    ANXIETY2   post-test anxiety
  by GROUP     exper-control grps

    Mean Rank     Cases

  ④   7.55          10  GROUP =  .00  intervention
     13.45          10  GROUP = 1.00  control
                    --
                    20  Total

                          Exact              Corrected for ties
  ⑤    U          W     2-Tailed P ⑥      Z        2-Tailed P
      20.5       75.5      .0232         -2.3331      .0196       ⑦
```

To determine whether the difference in the mean ranks for the two groups is statistically significant, the Mann-Whitney U statistics are used (⑤). The presented U value (20.5) represents the smaller of the two values, U_1 and U_2, and W (75.5) is the sum of the ranks for the group with the smaller number of subjects. When the group sizes are equal, W represents the sum of the ranks for the first presented level of the dichotomous variable (R_1). In our hypothetical study, W is the sum of the ranks for the intervention group (R_1) because it was assigned the value of 0 in the *Value Labels* subcommand in SPSS for Windows (i.e., 0 = intervention group, 1 = control group).

Because the total sum for any set of numbers from 1 to N is $[N(N+1)/2]$ (Siegel & Castellan, 1988), R_2 is obtained easily by subtracting R_1 from this total sum:

$$R_2 = \frac{N(N+1)}{2} - R_1 \quad \frac{20(20+1)}{2} - 75.5 = 210 - 75.5 = 134.5.$$

R_1 and R_2 can be obtained directly from the computer printout (Table 7.9) by multiplying the mean rank for a particular group by its sample size; for example, $13.45(10) = 134.5$.

Given this information, we can then reconstruct our values for U_1 and U_2 using the formulas outlined above and can see that $U = 20.5$ because that is the minimum of U_1 and U_2.

$$U_1 = (10)(10) = \frac{10(10+1)}{2} - 75. = 79.5$$

$$U_2 = (10)(10) = \frac{10(10+1)}{2} - 134.5 = 20.5.$$

In SPSS for Windows, the exact two-tailed significance level of U is presented when the sample size is less than 30 (⑥). For larger sample sizes, a two-tailed z statistic with corrections for ties is used (⑦). If the sample were large with quite a number of ties, we might want to use the results obtained from the z statistic. Because the sample size for this hypothetical study is small ($N = 20$) with relatively few ties in each group, we will use the exact two-tailed significance level of U to determine significance.

Because the alternative hypothesis is directional, it will be necessary to divide the two-tailed p value in half to determine whether this p value is less than the prestated alpha level. In our hypothetical study, the one-tailed p value ($.0232/2 = .0116$) is less than our $\alpha = .05$. The null hypothesis that the two groups come from the same population therefore is rejected. We conclude, therefore, that the intervention group had lower postintervention anxiety scores than the control group.

Presentation of Results

Table 7.10 is a suggested presentation of the results of the Mann-Whitney test. Note that we have presented the z statistic and the one-tailed p value for that z statistic ($p = .0196/2 = .0098$ rounded to .01). Because

Table 7.10 Suggested Presentation of Mann-Whitney Test Results

		Intervention			Control		Mann-Whitney	
	N	Mean	Median	N	Mean	Median	z	p
Postintervention anxiety	10	4.4	4.5	10	5.0	5.0	−2.33	.01

NOTE: The p value is for a one-tailed test.

the sample size is less than 30, it would also be possible to omit the z statistic and simply present the p value for the U statistic, as we did for the Fisher exact test (Table 7.4). These data could also be interpreted and presented in the text as follows:

> The results of the Mann-Whitney test indicate that the nurses evaluated the 10 children in the intervention group as having significantly lower postintervention anxiety than the 10 children in the control group ($z = -2.33$, one-tailed $p = .01$).

Advantages and Limitations of the Mann-Whitney Test

The Mann-Whitney test is a very powerful test that is sensitive to the central tendency of the scores. Relative to the independent t test, the Mann-Whitney test has excellent asymptotic efficiency, especially when the sample sizes are equal (Gibbons & Chakraborti, 1991; Hettmansperger, 1984). This means that it quickly approaches the power of the independent t test. Siegel and Castellan (1988) indicate that when the Mann-Whitney test is applied to data that meet the assumptions of the t test, the power efficiency of the Mann-Whitney test approaches 95.5% as N increases and is close to 95% for moderate size samples. This means that, even with moderate or small sample sizes, the Mann-Whitney test is nearly as powerful as the independent t test in being able to correctly reject the null hypothesis.

Parametric and Nonparametric Alternatives to the Mann-Whitney Test

The parametric counterpart to the Mann-Whitney test is the independent t test. As indicated, the independent t test is more powerful than the

Mann-Whitney test when the assumptions for this parametric test are met. When the assumptions are not met (e.g., the data are not distributed normally), the Mann-Whitney test is the more powerful test, especially when the tails of the distribution are heavy and there are outliers present in the data (Conover, 1980; Dexter, 1994; Zimmerman, 1993).

A nonparametric alternative that could be used if the shapes of the distributions are not similar is the median test (Chapter 8). When the distributions are similar, however, the median test is not as powerful as the Mann-Whitney test.

Examples From Published Research

Becker, P. T., Grunwald, P. C., Moorman, J., & Stuhr, S. (1993). Effects of developmental care on behavioral organization in very low birth weight infants. *Nursing Research, 42*, 214-220.

Hartgers, C., van Ameijden, E. J., van den Hoek, J. A., & Coutinho, R. A. (1992). Needle sharing and participation in the Amsterdam Syringe Exchange program among HIV seronegative injecting drug users. *Public Health Reports, 107*, 675-681.

McMurdo, M. E., & Rennie, L. M. (1994). Improvements in quadriceps strength with regular seated exercise in the institutionalized elderly. *Archives of Physical Medicine and Rehabilitation, 75*, 600-603.

Whiteman, K., Nachtmann, L., Kramer, D., Sereika, S., & Bierman, M. (1995). Effects of continuous lateral rotation therapy on pulmonary complications in liver transplant patients. *American Journal of Critical Care, 4*, 133-139.

Summary

In this chapter, we have examined three nonparametric tests that can be used when the independent variable is at the nominal level of measurement: the Fisher exact test, the chi-square test for two independent samples, and the Wilcoxon-Mann-Whitney U test (also called the Mann-Whitney test). We have also seen how another two-sample test, the Kolmogorov-Smirnov test, can be used to assess the similarity of shapes of distributions.

Both the Fisher exact test and the chi-square test for two independent samples are used when the dependent variable is categorical. The Fisher exact test is used with 2×2 contingency tables, whereas the chi-square test is useful for $2 \times k$ contingency tables. Neither of these two tests has a parametric alternative because the data being examined are categorical.

When the data are continuous, the Mann-Whitney test is preferred because it is sensitive to order in the data. This is a test for independent samples for which the dependent variable is at least at the ordinal level of measurement. The Mann-Whitney test is a very powerful test and is almost as powerful as its parametric counterpart, the independent t test. It is a particularly useful test when the data do not meet the assumptions of a t test, especially when the sample is small, the group sizes are unequal, and tails of the distributions are large. One of the assumptions of a Mann-Whitney test, however, is that the distributions of the two independent samples are similar in shape. When this assumption cannot be met, the median test can be used in its place.

The next chapter extends discussion to nonparametric tests that can be used when the independent variable has more than two levels. The four tests that will be examined are the chi-square test for n independent samples, the Mantel-Haenszel chi-square test for ordered categories, the median test, and the Kruskal-Wallis one-way ANOVA.

8 Assessing Differences Among Several Independent Groups

- **Chi-square test for k independent samples**
- **Mantel-Haenszel chi-square test for trends**
- **Median test**
- **Kruskal-Wallis one-way ANOVA test by ranks**

In Chapter 7, we examined nonparametric tests that would be suitable when the independent variable was dichotomous and the dependent variable was at any level of measurement. When the dependent variable is categorical, the Fisher's exact test or the chi-square test for two independent samples can be used. When it is continuous, the Mann-Whitney test is the test of choice.

In health care research, there are often more than two independent groups that are being compared. Subjects may be assigned randomly to one of three intervention conditions, or three patient groups having different diagnoses may be compared with regard to an outcome variable. A second independent categorical variable such as gender might also be included. If the dependent variable is continuous, the parametric tests that are commonly used to assess such data are one- and two-way ANOVAs (analyses of variance). A number of assumptions, however, are associated with the F statistics generated from these tests that potentially could be violated. For example, within each of the independent groups, there may be unequal cell sizes. Perhaps the dependent variable is not normally distributed or the variances are not equal. The dependent variable may also be at less than the interval level of measurement, or the researcher may want to generalize beyond a normally distributed population. If these are issues, the researcher may want to consider nonparametric alternatives to the available parametric tests.

In this chapter, we will examine nonparametric statistics for more than two independent groups. Four tests will be presented in detail: the chi-square test for k independent groups, the Mantel-Haenszel chi-square test for ordered categories, the median test, and the Kruskal-Wallis one-way ANOVA by ranks. We will also examine ways to assess strength of association for these tests and to determine the influences of different cells when faced with significant results.

The Chi-Square Test for
k Independent Samples

In Chapter 7, we examined a chi-square test for two independent samples. For this test, the independent variable was dichotomous and the dependent variable was categorical with two or more levels. The chi-square test for k independent samples is merely an extension of this test. It can be used when *both* the independent and dependent variables are categorical with two or more levels. This test also can be used when the researcher wants to examine the effects of several independent variables (e.g., gender and diagnosis) on a nominal-level outcome variable (e.g., posthospital adjustment: poor, moderate, or excellent). (Some researchers would argue that the two nominal-level variables, posthospital adjustment and social status, have order to their categories and therefore could be treated as ordinal-level variables. These particular variables have been deliberately selected for analysis in this example so that we can compare the results obtained from the chi-square test with those of the Mantel-Haenszel chi-square test. The latter test takes advantage of the ordered categories.)

An Appropriate Research Question
for the Chi-Square Test

The chi-square test for k independent samples (referred to herein as the *chi-square test*) is one of the most popular nonparametric tests. Pett, Lang, and Gander (1992) used the chi-square test to compare the effects of divorce in later life on changes in family rituals in three independent groups. Ejaz, Folmar, Kaufmann, Rose, and Goldman (1994) used this test to evaluate the effectiveness of a physical restraint reduction program among 144 frail elderly persons residing in a nursing home facility.

In our hypothetical study, we might be interested in assessing the relationship between family social status and mothers' assessments of their

Table 8.1 Example of Null and Alternative Hypotheses Suitable for Use
With the Chi-Square Test for *k* Independent Samples

Null Hypothesis

H_0:   There is no association between a family's social status (e.g., low, medium, or high)
and a mother's evaluation for her child's posthospital adjustment (e.g., poor,
moderate, or excellent); that is, the two variables, family social status and children's
posthospital adjustment, are independent.

Alternative Hypothesis

H_a: There is an association between a family's social status (e.g., low, medium, or high)
and a mother's evaluation for her child's posthospital adjustment (e.g., poor,
moderate, or excellent); that is, the two variables, family social status and children's
posthospital adjustment, are dependent.

---

children's posthospital adjustment. Suppose that the social status of the
families in our study was assessed on a nominal level of measurement: low,
medium, or high social status. (Once again, it could be argued that this
variable is measured at the ordinal level; it was selected so that results from
this test could be compared to those of the Mantel-Haenszel chi-square
test.) Mothers were also asked 2 weeks following their children's hospi-
talization to assess the children's posthospital adjustment on a scale of 1 to
100. This scale was then collapsed, using prespecified criteria, into three
categories: *poor* (scores ≤ 70), *moderate* ($71 \leq$ scores ≤ 85), and *excellent
adjustment* (scores >85). (It is recognized that most researchers would
prefer not to collapse an interval-level variable. This example is used to
compare the results from several nonparametric tests.) Given this collapsed
categorical outcome variable, the following research question would be
appropriate for use with a chi-square test:

What is the relationship between a family's social status (low, medium, or high)
and a mother's assessment of her child's posthospital adjustment (poor, moderate, or excellent)?

Null and Alternative Hypotheses

Table 8.1 presents examples of possible null and alternative hypotheses
generated from our research question that would be suitable for use with a
chi-square test. Note that, like the chi-square test for two independent

samples (Chapter 7), these hypotheses address independence of variables and are nondirectional.

Overview of the Procedure

The chi-square test for k independent samples proceeds in the same way as the chi-square two-sample test from Chapter 7. Because the procedure was outlined in detail in that chapter, we will review it only briefly here.

First, the frequency data are placed into an $r \times c$ contingency table where $r =$ the number of levels of the independent variable and $c =$ the number of levels of the dependent variable. In our hypothetical example, because there are three levels of both social status and posthospital adjustment, ours would be a 3×3 contingency table. Next, using the formulas outlined in Chapter 7, the number of cases that would be expected in each of the cells if the null hypothesis of independence of variables were true is calculated. From this data, a chi-square statistic (χ^2) is generated by comparing the observed frequencies with the expected frequencies in each cell using the following formula:

$$\chi^2 = \sum_{i=1}^{r} \sum_{j=1}^{c} \frac{(O_{ij} - E_{ij})^2}{E_{ij}}$$

where

O_{ij} = the observed number of cases for a cell in the ith row and jth column

E_{ij} = the expected number of cases for the same cell if the null hypothesis were true

$\Sigma\Sigma$ = a double summation of the fraction across all rows (r) and columns (c).

If the data meet the assumptions of the chi-square test, this generated χ^2 statistic is distributed as a χ^2 with $(r-1)(c-1)$ degrees of freedom. The null hypothesis of independence of variables is rejected in favor of the alternative hypothesis if the generated χ^2 is greater than the critical value of the χ^2 at the specified alpha level and designated number of degrees of freedom. Alternatively, in examination of a computer printout, the null hypothesis is rejected if the generated significance level (p) is less than alpha (e.g., $\alpha = .05$).

Critical Assumptions of the Chi-Square Test

The critical assumptions of the chi-square test for k independent samples are similar to those for the chi-square test for two independent samples.

1. *Both the independent and dependent variables are categorical.*
2. *The data consists of frequencies, not scores.*
3. *Each randomly selected observation can be classified into only one category for the independent variable(s) and only one category for the dependent variable. There are no repeated observations and no multiple response categories.*

Our data partially meet all three of these assumptions because the two variables, social status and posthospital adjustment, are categorical and the data consist of frequencies that occupy mutually exclusive categories. The data, however, have not been randomly selected.

4. *No more than 20% of the cells may have expected frequencies of less than 5, and no cell should have an expected frequency of less than 1.*

The rule given in Assumption 4 is particularly important for a contingency table that is larger than 2 × 2. For example, in order for a 3 × 3 table to meet this rule, the total sample size must be at least 45 subjects if the marginal totals for all levels of the independent and dependent variables are equal.

In our hypothetical study, we started out with only 20 subjects. With such a small sample size, we would clearly fail to meet this important rule. If the data do not meet this rule, a possible solution would be to create a smaller table by combining the cells of the dependent and/or independent variables. For example, we might collapse our categorical variables to *low* versus *nonlow* social status and for *poor* versus *good* adjustment. The decision to collapse certain cells must be undertaken carefully so that the combined cells make intuitive sense given the research questions of the study.

Because this is a hypothetical study, we have the liberty of increasing our sample size without worrying about its consequences (e.g., cost and feasibility). Suppose there were 60 instead of only 20 children in the study.

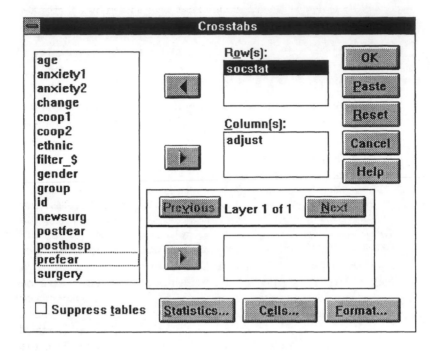

Figure 8.1. SPSS for Windows Commands for the Chi-Square Test for *k*
Independent Samples

This larger sample size increases the likelihood that the rule in Assumption
4 has been met. This assumption will be assessed in detail when we examine
the computer printouts.

Computer Commands

Figure 8.1 presents the SPSS for Windows commands that are used to
generate the chi-square test for *k* independent samples. This dialogue box
is opened by clicking on *Statistics . . . Summarize . . . Crosstabs* in the
menu. To obtain the chi-square test, the *Statistics* box is opened and the
desired statistics are selected (e.g., *Chi-Square tests* to assess independence
and *Cramér's V* to measure the strength of association). Additional cell
information (e.g., observed and expected frequencies, row and column

percentages, and standardized residuals) can be obtained by opening the *Cells* dialogue box.

Note in Figure 8.1 that the row variable in this hypothetical example is the independent variable, family social status (*socstat*), whereas the column variable is the dependent variable, posthospital adjustment (*adjust*). A second independent variable, such as gender, could have been selected by clicking on the variable *gender* and moving it into the *Layer* box. It is important, however, to be parsimonious regarding layers, because each additional independent variable will increase sample size requirements.

Computer-Generated Output

Table 8.2 presents the syntax commands and computer-generated printout from SPSS for Windows for the chi-square test for the 60 children in our hypothetical study. Because our syntax commands (①) were nearly identical to those for the chi-square test for two independent samples, this printout is very similar to that presented in Chapter 7.

First, we are given the positioning of the cell information that we requested (②). The first value for each cell represents the actual frequency (*Count*). The second value represents the expected frequency for the cell under the null hypothesis of independence of variables (*Exp Val*). The third value is the percentage expressed in terms of row totals (*Row Pct*). The fourth value is the percentage expressed in terms of column totals (*Col Pct*), and the final value in each cell represents the standardized residuals (*Std Res*). For example, in row 2, column 3, there are four children (③) of middle social status whose mothers rated them as having "excellent" posthospital adjustment. The expected value for any cell is the product of the cell's row and column totals divided by the total sample size. For the cell that occupies row 2, column 3, we would have expected $(25)(17)/60 = 7.1$ children (④) if the null hypothesis of independence of variables were true. The difference between the actual frequency and the expected value suggests that the two variables, social status and posthospital adjustment, may not be independent.

Additional cell information can be gleaned from the printout. For example, looking at the row percentage (*Row Pct*) for that same cell (⑤), 16.0% of the 25 children who were of middle social status were rated by their mothers to have "excellent" posthospital adjustment. Concerning the column percentages (*Col Pct*) for that cell (⑥), 23.5% of the 17 children who were rated by their mothers as having "excellent" posthospital adjustment were of middle social status.

Table 8.2 Computer-Generated Printout Obtained From SPSS for
            Windows for the Chi-Square Test for k Independent Samples

```
CROSSTABS
  /TABLES=socstat  BY adjust
  /FORMAT= AVALUE NOINDEX BOX LABELS TABLES
  /STATISTIC=CHISQ PHI    ①
  /CELLS= COUNT EXPECTED ROW COLUMN SRESID    ②

SOCSTAT  social status  by  ADJUST  posthospital adjustment

                    ADJUST                Page 1 of 1
             Count  |
             Exp Val |
     ②      Row Pct |Poor    Moderate Excellent
             Col Pct |                             Row
             Std Res |   1.00|    2.00|    3.00|  Total
SOCSTAT    --------+--------+--------+--------+
             1.00  |    6  |    7  |    3  |    16
  low              |   4.3 ⑧|  7.2  |  4.5 ⑨|  26.7%
                   |  37.5% | 43.8% | 18.8% |
                   |  37.5% | 25.9% | 17.6% |
                   |   .8   |  -.1  |  -.7  |
                   +--------+--------+--------+
             2.00  |    6  |   15  |    4 ③|    25
  middle           |   6.7 | 11.3  |  7.1 ④|  41.7%
                   |  24.0% | 60.0% | 16.0% ⑤
                   |  37.5% | 55.6% | 23.5% ⑥
                   |  -.3  |  1.1  | -1.2  |
                   +--------+--------+--------+
             3.00  |    4  |    5  |   10  |    19
  high             |   5.1 |  8.6  |  5.4  |  31.7%
                   |  21.1% | 26.3% | 52.6% |
                   |  25.0% | 18.5% | 58.8% |
                   |  -.5  | -1.2  |  2.0 ⑬
                   +--------+--------+--------+
          Column      16      27      17      60
          Total     26.7%   45.0%   28.3%   100.0%

     Chi-Square                   Value          DF    Significance
--------------------            ----------       ----  ------------
Pearson                         9.54490 ⑩         4     .04883 ⑪
Likelihood Ratio                9.14369           4     .05761
Mantel-Haenszel test for        4.11150 ⑭         1     .04259 ⑯
     linear association

Minimum Expected Frequency -     4.267
Cells with Expected Frequency < 5 -      2 OF    9 ( 22.2%) ⑦

Approximate
    Statistic               Value      ASE1   Val/ASE0 Significance
--------------------       ---------  -------- -------- ------------
Phi                         .39885                        .04883
Cramer's V                  .28203 ⑫                      .04883
Pearson's R                 .26400 ⑮  .13152   2.08436    .04154
```

We can now examine the printout to determine the extent to which we have met Assumption 4: No more than 20% of the cells may have expected frequencies less than 5. The computer printout in Table 8.2 indicates that only 2 of 9 cells (22.2%) have expected frequencies of less than 5 (⑦). The problem cells are those families of lower social status whose mothers either rated their children as having "poor" adjustment (*Exp Val* = 4.3, ⑧) or "excellent" adjustment (*Exp Val* = 4.5, ⑨). Given that we have nearly met the requirement of Assumption 4 and that there is no clear way to collapse these data to eliminate the slight discrepancy, we will proceed with the test.

The generated χ^2 statistic can now be examined to determine whether the null hypothesis of independence of variables should be rejected. Our decision rule is to reject the null hypothesis if the generated χ^2 value exceeds the critical value at our specified α level with $df = (r - 1)(c - 1)$, or, synonymously, if the generated significance level is less than the prestated alpha level. In our hypothetical study, $\alpha = .05$ and $df = 4$ ([$r - 1$][$c - 1$] = [3 - 1][3 - 1] = 4); according to a table of critical χ^2 values (Hinkle et al., 1994), $\chi^2_{cv} = 9.49$.

Our calculated χ^2 value is 9.54 (⑩), which is greater than the critical value of 9.49. The two-tailed p value for this calculated χ^2 statistic, $p = .048$ (⑪), is less than $\alpha = .05$. The conclusion to be reached, therefore, is that there is a significant association between families' social status and mothers' evaluations of their children's posthospital adjustment; that is, the two variables, family social status and children's posthospital adjustment, are dependent.

Determining the Strength and Meaning
of a Significant Association

The chi-square test is nondirectional. It tells us whether or not the variables under consideration are associated but tells us nothing about the strength of the association or which cells appear to be influencing the outcome observed. The Cramér's *V* statistic (Chapter 7) enables the reader to determine the strength of the association between the variables. There are two other useful approaches to ascertaining which cells have the most impact on the outcome: (1) examination of the standardized residuals (Hinkle et al., 1994) and (2) partitioning the $r \times c$ contingency table into smaller 2 × 2 subtables to determine where the differences are occurring (Siegel & Castellan, 1988).

Determining Strength of Association

Once a significant association has been found between two categorical variables via the chi-square test, the next task is to determine the strength of this relationship. The computer printout in Table 8.2 reports two measures of association: the phi coefficient and Cramér's V statistic. The phi coefficient is used for 2 × 2 tables, and Cramér's V statistic is used for contingency tables that are larger than 2 × 2. As indicated in Chapter 7, the formula for Cramér's V statistic is:

$$V = \sqrt{\chi^2/N(L-1)}$$

where

χ^2 = the generated chi-square statistic

N = the total number of cases

L = the minimum number of rows or columns in the contingency table.

This Cramér statistic ranges from 0 to 1, where 0 implies no relationship and 1 a perfect relationship. Its significance level will be the same as for the χ^2 statistic.

The computer printout in Table 8.2 indicates that Cramér's V = .282 (⑫). That is,

$$V = \sqrt{\chi^2/N(L-1)} = \sqrt{9.54/60\,(3-1)} = .282$$

Using the guidelines outlined in Chapter 7 (Table 7.6) for the strength of association for the phi coefficient, we would conclude that the strength of the relationship between social status and posthospital adjustment is weak.

Determining Cell Influences: Examining Residuals

There are several approaches to determining which cells most strongly influence the χ^2 value. One of the easiest approaches is to examine the sizes of the standardized residuals for each cell (see Chapter 7 for details). Cells that have standardized residuals greater than 2 in absolute value are categories that are major contributors to the significant χ^2 value (Hinkle et al., 1994).

In Table 8.2, there is only one cell that has a standardized residual that comes close to meeting this criterion: those families of high social status in which mothers rated their children as having "excellent" posthospital adjustment (standardized residual = 2.0, ⑬). Comparing the actual and expected frequencies for that cell, more mothers of high social status rated their children as having excellent posthospital adjustment ($n = 10$) than would have been expected by chance ($n = 5.4$).

Partitioning the Contingency Table

A second approach to examining cell influences is to subpartition the contingency table into a series of 2 × 2 independent subtables (Siegel & Castellan, 1988). The degrees of freedom available for the chi-square analysis dictate how many tables can be set up: There can be one independent 2 × 2 table for each degree of freedom (i.e., 4 df = 4 tables). Because of the limited number of comparisons that can be made, the researcher needs to identify carefully which comparisons are most meaningful to the study.

Although the partitioning approach may initially appear confusing and cumbersome, it is really quite straightforward. First, the contingency table is arranged in such a way that the most meaningful 2 × 2 comparisons of the categorical data are made given the research questions of the study (Siegel & Castellan, 1988). Because the variables being examined are nominal, there is no restriction as to the order in which the rows and columns are placed in the table.

Siegel and Castellan indicate that, for clarity of understanding, it is best to begin the 2 × 2 partitioning in the upper lefthand corner of the table and then collapse the rows and columns of the table in a systematic and meaningful fashion. Figure 8.2 outlines the procedure for collapsing a 3 × 3 table into four independent 2 × 2 tables. There are four 2 × 2 tables because, for a 3 × 3 table, there are 4 degrees of freedom ($df = [r - 1][c - 1] = [3 - 1][3 - 1] = 2$).

The lower righthand cell of each 2 × 2 subtable to be created always consists of a single set of frequencies and serves as the foundation for the 2 × 2 partition (Figure 8.2a-d, ①). The upper lefthand cell in the 2 × 2 table consists of the sum of all frequencies that lie directly up and to the left of this foundation cell (②). The lower left cell of the 2 × 2 subtable consists

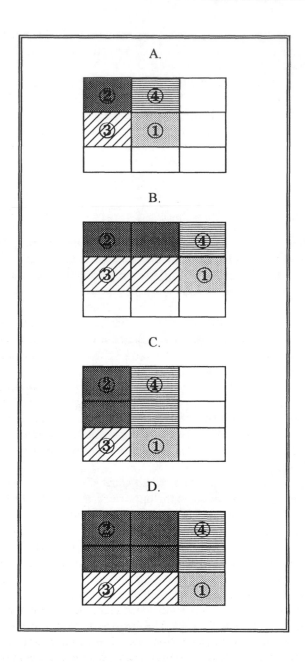

Figure 8.2. Partitioning a 3 × 3 Contingency Table

of the sum of all frequencies that lie directly to the left of the foundation cell (③), whereas the upper right cell is the sum of the frequency values that lie directly above the foundation cell (④).

Following this pattern, the first partition consists of the four adjacent cells in the upper lefthand corner of the 3 × 3 table (Figure 8.2a). The second partition (Figure 8.2b) is begun by moving one cell to the right of the foundation cell identified in the first partition (①). The sum of all frequencies that lie in the cells directly above and to the left of this new foundation cell make up cell ② in the second 2 × 2 subtable. Cell ③ consists of the sum of the frequencies in the cells that lie directly to the left of cell ①, and cell ④ is made up of all frequencies that lie directly above cell ①.

The third partition (Figure 8.2c) is obtained by targeting the single cell directly below the foundation cell identified in the first partition as the new cell ①. Following the guidelines outlined above, cells ②, ③, and ④ can be obtained easily. A similar approach is taken for the fourth and final partition of the 3 × 3 table (Figure 8.2d), except that, now, cell ① consists of the single cell that lies directly to the right of foundation cell ① in the third partition.

Each of these four 2 × 2 subpartitions can now be treated as an independent 2 × 2 table, and chi-square statistics can be calculated for each table. The resulting χ^2 values that are generated from the four 2 × 2 comparisons sum to the original χ^2 value.

The calculations of the four chi-square statistics are tedious to perform by hand. Unfortunately, SPSS for Windows does not have the capability of partitioning contingency tables, but fortunately, Siegel and Castellan (1988) present an interactive computer program designed to undertake such analyses. This program, written in BASIC, can be used on any Windows- or DOS-supported computer system that has the capacity for programming in BASIC. This partitioning program has been reproduced and rewritten for QBASIC (see Appendix) with permission from the publishers of the Siegel and Castellan text. It will be used to partition the 3 × 3 table for our hypothetical study.

Using the approach outlined above, Table 8.3 presents the partitioned 2 × 2 subtables for the 3 × 3 contingency table that examines the relationship between social status and posthospital adjustment. Note that 2 × 2 subpartition 1 compares *Low* and *Middle* social status families with regard to *Poor* or *Moderate* posthospital adjustment. Subpartition 2 compares *High* and *Not High* social status families with regard to *Poor* or *Moderate* posthospital adjustment. Subpartition 3 examines *Low* and *Middle* social status families with regard to *Excellent* or *Not Excellent* posthospital adjustment. Finally, subpartition 4 makes the 2 × 2 comparison of *High* and

Table 8.3 Partitioning the 3 × 3 Contingency Table Into Four
Independent 2 × 2 Tables

a.   Subpartition #1:   Social Status (Low, Middle) vs. Posthospital Adjustment
      (Poor, Moderate)

Social Status	Posthospital adjustment			Total
	Poor	Moderate	Excellent	
Low	6	7	3	16
Middle	6	15	4	25
High	4	5	10	19
Total	16	27	17	60

b.   Subpartition #2: Social Status (High, Not High) vs. Posthospital Adjustment
      (Poor, Moderate)

Social Status	Posthospital adjustment			Total
	Poor	Moderate	Excellent	
Low	6	7	3	16
Middle	6	15	4	25
High	4	5	10	19
Total	16	27	17	60

c.   Subpartition #3: Social Status (Low, Middle) vs. Posthospital Adjustment
      (Excellent, Not Excellent)

Social Status	Posthospital adjustment			Total
	Poor	Moderate	Excellent	
Low	6	7	3	16
Middle	6	6	4	25
High	4	5	10	19
Total	16	27	17	60

d.   Subpartion #4: Social Status (High, Not High) vs. Posthospital Adjustment
      (Excellent, Not Excellent)

Social Status	Poor	Moderate	Excellent	Total
Low	6	7	3	16
Middle	6	15	4	25
High	4	5	10	19
Total	16	27	17	60

Not High social status regarding *Excellent* or *Not Excellent* posthospital adjustment.

Statistically Analyzing the Partitioned Subtables

To undertake the statistical analyses of these four 2 × 2 tables, the interactive computer program written by Siegel and Castellan (1988) (see Appendix) needs to be written as a DOS textfile and run as a computer program in BASIC (e.g., we used QBASIC, DOS V5.0). All lines should be typed carefully as presented in the Appendix. Do not change the upper- and lowercase letters. In QBASIC, this program is started by typing QBASIC /RUN [*name of file written in BASIC*] at the DOS prompt and completing the answers that are asked about the contingency table to be analyzed. Table 8.4 presents the results of partitioning the contingency table from our hypothetical study using the QBASIC program from the Appendix.

As Table 8.4 indicates, the program initially asks you to specify the number of rows and columns in your contingency table (①). In our example, we indicated that there were 3 rows and 3 columns. Next, the program requests the frequencies for each cell, starting with row 1, column 1 (②). This request continues until all the cells for the $r \times k$ contingency table have been specified.

From these data, a χ^2 output is generated on the computer screen (Table 8.4). This result indicates that, thank goodness, we have obtained the same χ^2 value as in Table 8.2 (9.54, ⑩). Now, however, the χ^2 value has been broken down into four subpartitions, each reflecting the comparisons made in Table 8.3a-d (④). Because these tests are independent, the χ^2 values for all four partitions will sum to 9.54. Note that each partition is identified by its foundation cell, the cell in the bottom right corner of the 2 × 2 partition. The foundation cell for the first partition, for example, is cell 2:2 (⑤), or that cell occupying row 2, column 2 in the 3 × 3 table being analyzed. These four subpartitions are thus presented in the same order as the partitions outlined in Table 8.3a-d.

To determine which of these four 2 × 2 partitions are significant, each generated χ^2 value needs to be compared to a critical χ^2 value for a 2 × 2 table under the null hypothesis of independence of variables. Table 8.4 indicates that the critical χ^2 value for a 2 × 2 table with $df = 1$ at $\alpha = .05$ is 3.84 (⑥). In our hypothetical study in which $\alpha = .05$, our decision rule is that we will reject the null hypothesis of independence of variables if the generated χ^2 value is larger than 3.84.

Table 8.4 Results of Partitioning the Contingency Table Using the Basic Program From the Appendix

```
PARTITIONING AN r x k CONTINGENCY TABLE INTO 2 x 2 INDEPENDENT SUBTABLES

Originally presented in 1988 by N. J. Castellan, Jr.
Updated for QBASIC by Dr. M. Pett in 1996
Reprinted with permission from McGraw-Hill Publishers.

You must enter the size of the contingency table.

How many rows does your contingency table have? 3
How many columns does it have? 3  ①

Please enter the data cell by cell when requested.  ②
In addition, press the ENTER key after each ?

Enter the data for the cell that occupies Row 1 Column 1 ?  6
Enter the data for the cell that occupies Row 1 Column 2 ?  7
Enter the data for the cell that occupies Row 1 Column 3 ?  3
Enter the data for the cell that occupies Row 2 Column 1 ?  6
Enter the data for the cell that occupies Row 2 Column 2 ? 15
Enter the data for the cell that occupies Row 2 Column 3 ?  4
Enter the data for the cell that occupies Row 3 Column 1 ?  4
Enter the data for the cell that occupies Row 3 Column 2 ?  5
Enter the data for the cell that occupies Row 3 Column 3 ? 10

Chi-square = 9.544899  with  4  degrees of freedom.  ③

Partition cell(i,j)       Chi-Square  ④
      1           2: 2 ⑤      1.229
      2           3: 2        0.195
      3           2: 3        0.036
      4           3: 3        8.085  ⑦

The critical Chi-Square value for df=1, alpha=.05 is 3.84  ⑥
The critical Chi-Square value for df=1, alpha=.01 is 6.64
```

The only χ^2 value in Table 8.4 that exceeds the critical value is that of partition 4: 8.085 (⑦). To interpret the meaning of this significance, we need to return to Table 8.3d. This partition examined the association between the *High* and *Not High* social status families with regard to their *Excellent* or *Not Excellent* posthospital adjustment. The significant result confirms what we already suspected from examining the standardized residuals: Mothers from families with higher social status were significantly more likely to rate their children's posthospital adjustment as

Table 8.5 Presentation of the Chi-Square Test Analyses of Children's Posthospital Adjustment in Table Form

Social Status	Poor		Moderate		Excellent		Total		χ^2	p
	N	Percentage	N	Percentage	N	Percentage	N	Percentage		
Low	6	37.5	7	25.9	3	17.7	16	26.6	9.54	.049
Middle	6	37.5	15	55.6	4	23.5	25	41.7		
High	4	25.0	5	18.5	10	58.8	19	31.7		
Total	16	100.0	27	100.0	17	100.0	60	100.0		

"excellent" than were mothers from families with lower or middle social status.

Presentation of Results

Table 8.5 presents the results of the chi-square test for k independent samples in tabular form. These results also could be interpreted and presented in the text as follows:

> The results of the chi-square test indicated a significant but weak association between family social status and posthospital adjustment ($\chi^2 = 9.54$, $p = .049$, Cramér's $V = .28$). Examination of the standardized residuals and subpartitioning of the contingency table (Siegel & Castellan, 1988) revealed that mothers of higher social status were significantly more likely than other mothers to rate their children's posthospital adjustment as "excellent."

Advantages and Limitations of the Chi-Square Test

The chi-square test is an extremely useful and flexible technique to be used when both the independent and dependent variables are categorical. Because it is insensitive to order, the chi-square test is not the test of choice when either of the categorical variables has ordered levels. It is also improper to use this test when the data are not independent, for example, when there are multiple response categories, there are repeated observations, or the data are matched or paired samples.

Parametric and Nonparametric Alternatives to the Chi-Square Test

There are no parametric alternatives to the chi-square test because it is intended for use with categorical data. When the data do not meet the assumptions of the chi-square test for k independent samples, the data could be collapsed into smaller contingency tables for use with the Fisher exact test or the chi-square test for two independent samples (Chapter 7). If either of the categorical variables has ordered levels, alternative nonparametric tests are the other tests reviewed in this chapter: the Mantel-Haenszel chi-square test for trends, the median test, and the Kruskal-Wallis one-way ANOVA by ranks.

When matched or paired samples are to be compared with regard to the relative risk associated with a nominal-level categorical variable, the Mantel-Haenszel odds ratio (OR) test is a useful statistic (Rothman, 1986). When more than two categorical independent variables are being examined for their influence on a categorical dependent variable, two potentially useful approaches to data analysis are logistic regression and discriminant analysis (Munro & Page, 1993).

Examples From Published Research

Ejaz, F. K., Folmar, S. J., Kaufmann, M., Rose, M. S., & Goldman, B. (1994). Restrain reduction: Can it be achieved? *The Gerontologist, 34*, 694-699.
Pett, M. A., Lang, N., & Gander, A. (1992). Late life divorce: Its impact on family rituals. *Journal of Family Issues, 13*, 526-532.

The Mantel-Haenszel Chi-Square Test for Trends

It often happens in health care research that the categorical variables being examined have order to their levels (e.g., stages of disease, length of exposure, age levels, or categories of social status) yet are not sufficiently ordinal to be considered continuous variables. Use of the chi-square test for k independent samples with these data would result in loss of potentially valuable information because this test is insensitive to order. There is an extension of the chi-square test, the Mantel-Haenszel (M-H) chi-square test for trends, that takes order into account (Mantel, 1963; Mantel & Haenszel,

1959). It is, therefore, a useful test when the values assigned to the categories of both independent and dependent variables are ordered.

The M-H test has been used widely in epidemiologic case control studies in which there are stratified 2 × 2 tables and the researcher is interested in comparing the likelihood of the occurrence of an event (e.g., contracting cancer) given two groups who have been exposed versus unexposed to a risk factor (e.g., cigarette smoking) and who have been matched or paired with regard to certain characteristics (e.g., gender or age) (Armitage & Berry, 1994; Kahn & Sempos, 1989; Rothman, 1986). It has also been used extensively in educational testing to examine differences in individual performance and group ability on a test between two comparable groups of test takers (e.g., males and females) (Holland & Wainer, 1993). The focus of this presentation will be on the use of the M-H chi-square test to examine trends in data when at least one of the two categorical ordered variables has more than two levels and the data are not matched or paired.

An Appropriate Research Question for the
Mantel-Haenszel Chi-Square Test

Although the M-H test for stratified case control data has been used extensively in the epidemiology and medical literature, the M-H test for trends has been used sparingly. Glazer, Morgenstern, and Doucette (1994) used the M-H test for trends to examine psychosocial, clinical, and medical correlates of race that might help to explain an observed association between race and incidence of tardive dyskinesia among a group of 398 seriously mentally ill outpatients. Vlahov, Anthony, Celentano, Solomon, and Chowdhury (1991) used the M-H test to assess changes in injection practices among 421 intravenous drug users during the period 1982-1987. Wright and Carlquist (1994) used the same test to examine age-related differences of measles immunity in 599 employees working in a large hospital setting, and Janicak and colleagues (1985) used the M-H procedure in their meta-analysis of the efficacy of ECT compared with other interventions in the treatment of severe depression.

In our hypothetical intervention study, there are several areas in which the M-H test could be used. In the previous example, we used the chi-square test for k independent samples to examine the association between two categorical variables, social status and posthospital adjustment. Note that each of these variables had three levels that were ordered: The social status variable had three levels of low, medium, and high social status, and the posthospital adjustment variable had three levels of poor, moderate, and excellent posthospital adjustment.

Table 8.6 Example of Null and Alternative Hypotheses Suitable for Use
With the Mantel-Haenszel Chi-Square Test

Null Hypothesis

H_0:   There is no linear trend associated with a family's social status (low, medium, or
high) and a mother's evaluation for her child's posthospital adjustment (poor,
moderate, or excellent).

Alternative Hypothesis

H_a: There is a linear trend associated with a family's social status (low, medium, or high)
and a mother's evaluation for her child's posthospital adjustment (poor, moderate, or
excellent).

The chi-square test that was used to determine whether there was an
association between these two variables was restricted in its ability to
detect the degree of linearity between the two variables. The M-H chi-
square test for trends could have been used as an alternative. A research
question that could be answered using this test would be as follows:

Is there a linear trend associated with a family's level of social status (low,
medium, or high) and a mother's evaluation of her child's posthospital adjustment (poor, moderate, or excellent)?

Null and Alternative Hypotheses

Table 8.6 presents an example of null and alternative hypotheses generated from our research question that would be suitable for use with an M-H
chi-square test. Like the ordinary chi-square test, the M-H test is a nondirectional two-tailed test of significance.

Overview of the Procedure

The data to be analyzed are first cast into an $r \times c$ contingency table in
which there are two categorical variables whose levels are ordered. Unlike
in the ordinary chi-square test, the order of placement of categories now is
important because this is a test of linear trends. The M-H chi-square test
cannot be used with categorical data whose levels are not ordered. Rothman
(1986) suggests that should the researcher suspect a nonlinear trend a
priori, the categories can be reassigned category values based on the
hypothesized trend.

The M-H chi-square test is cumbersome to calculate by hand but is readily available in statistical computer packages. For the exact computational formula, the interested reader is referred to Armitage and Berry (1994) and Rothman (1986). Conceptually, like the ordinary chi-square test, the M-H test for trends compares the actual frequencies in each cell with what would have been expected under the null hypothesis of no linear trend among the ordered categories. If the data meet the assumptions for a M-H chi-square test, this χ^2_{MH} is distributed as a χ^2 with $df = 1$. This χ^2_{MH} has an easier formula (Rothman, 1986):

$$\chi^2_{MH} = \frac{[\sum(O_{ij} - E_{ij})]^2}{\sum \text{Var}(O_{ij} - E_{ij})} = r^2 [n - 1)$$

where

r^2 = the square of the Pearson product-moment correlation

n = the total sample size.

The decision rule will be to reject the null hypothesis of no linear trend in the ordered categories if the calculated χ^2_{MH} is greater than a critical value of the χ^2 with $df = 1$ at the specified alpha level (e.g., if $\alpha = .05$, $\chi^2_{cv} = 3.84$).

**Critical Assumptions of the
Mantel-Haenszel Chi-Square Test**

The M-H chi-square test shares some of the assumptions of the ordinary chi-square test and has additional assumptions of its own.

1. *Both independent and dependent variables are categorical, with two or more levels.*
2. *The levels of both categorical variables are ordered.*
3. *The data consist of frequencies, not scores.*
4. *Each randomly selected observation can be classified into only one category for the independent variable(s) and only one category for the dependent variable. There are no repeated observations and no multiple response categories.*

Although adequate sample size has been alluded to with regard to the M-H test when used with case control data (Kahn & Sempos, 1989;

Rothman, 1986), no direct reference has been found for adequate sample size when examining descriptive data for trends. Because this is a chi-square test with $df = 1$, it would be reasonable to assume that the M-H test shares the following chi-square test requirement:

> 5. *No more than 20% of the cells may have expected frequencies of less than 5, and no cell should have an expected frequency of less than 1.*

The variables of interest in our hypothetical intervention study (social status and posthospital adjustment) meet most of these requirements. The frequency data that have been generated are at the nominal level, ordered, with mutually exclusive categories. There are some problems, however. The data have not been randomly selected, and slightly more than 20% of the cells (22.2%) have expected frequencies less than 5. This second problem, expected cell frequencies, is not so bad, however, as to prevent our continuing with the analysis.

Computer Commands

The computer commands to obtain the M-H chi-square test are quite straightforward and are similar to those for the ordinary chi-square test: *Statistics . . . Summarize . . . Crosstabs*. In SPSS for Windows, the researcher is given the results of this test whenever *Chi-Square* statistics are requested from the *Statistics* dialogue box (Figure 8.1).

Computer-Generated Output

Because the commands for the M-H test for trends are identical to those of the ordinary chi-square test, the syntax commands and output that are generated are similar (Table 8.2). The M-H test for linear trends (χ^2_{MH} = 4.1115) is presented under the chi-square tests (Table 8.2, ⑭). Because the Pearson correlation (r) between social status position and posthospital adjustment is .2640 (⑮), the χ^2_{MH} could be calculated directly:

$$\chi^2_{MH} = r^2(n - 1) = .2640^2(60 - 1) = 4.112,$$

which is similar to the rounded χ^2_{MH} value presented in Table 8.2. Is this sufficiently large to reject the null hypothesis given our two-tailed $\alpha = .05$? To make that decision, we could compare our actual $\chi^2_{MH} = 4.1115$ with $\chi^2_{cv} = 3.84$. Because the actual value is greater than the critical value, we can reject the null hypothesis. Because we have the computer printout, we can see that the significance value (p) for this χ^2_{MH} is .043 (⑯), which is less than $\alpha = .05$, again indicating that the null hypothesis is to be rejected. The conclusion to be drawn, therefore, is that there is a significant linear trend between social status and posthospital adjustment: As family social status increases, mothers tend to rate their children's posthospital adjustment higher.

Is this linear trend very strong? To determine the strength of this trend, we would look at Cramér's V (.282, ⑫), which we already have determined to be of low strength. The weakness of this linear trend is further supported by the lack of noticeable change in the significance level from the ordinary chi-square test ($p = .049$, ⑪) to the Mantel-Haenszel chi-square test of trends ($p = .043$, ⑯). If there had been a stronger linear trend, a more dramatic change in the p value would have occurred.

Presentation of Results

The results of the Mantel-Haenszel test for trends could be presented in a manner similar to that for the ordinary chi-square test (Table 8.5) with the χ^2_{MH} presented instead of the χ^2 value. These results also could be stated in the text:

> The results of the Mantel-Haenszel chi-square test for trends indicated a significant but weak linear trend between family social status and posthospital adjustment ($\chi^2_{MH} = 4.11$, $p = .043$, Cramér's $V = .28$). Examination of the standardized residuals and subpartitioning of the contingency table (Siegel & Castellan, 1988) revealed that mothers of higher social status were significantly more likely than other mothers to rate their children's posthospital adjustment as "excellent."

Advantages and Limitations of the Statistic

The Mantel-Haenszel chi-square test for trends is a somewhat misunderstood and underused chi-square test that is very useful when the categorical variables to be examined have ordered levels. Because of its relationship to the Pearson r, the Mantel-Haenszel chi-square test for trends is sensitive to order and, unlike the ordinary chi-square test, is able to detect linear trends among ordered nominal-level variables. This makes the Mantel-Haenszel chi-square test a more powerful alternative to the ordinary chi-square test when the levels are ordered. It should not be used, however, when the levels of the nominal-level variable do not have order or when the basic assumptions of the ordinary chi-square test have been violated.

Parametric and Nonparametric Alternatives to the Mantel-Haenszel Chi-Square Test

Because the variables that are being examined with the Mantel-Haenszel chi-square test for trends are categorical, there is no parametric equivalent to this test. As with the ordinary chi-square test, the data can be collapsed into a smaller table should the sample size not be large enough to meet this test's assumptions. If both sets of data to be examined are ordinal, a more sensitive alternative to this test is the Spearman rho. If one variable is categorical and the other continuous, the Kruskal-Wallis test would be an alternative nonparametric test to consider.

Examples From Published Research

Glazer, W. M., Morgenstern, H., & Doucette, J. (1994). Race and tardive dyskinesia among outpatients at a CMHC. *Hospital and Community Psychiatry, 45,* 38-42.

Janicak, P. G., Davis, J. M., Gibbons, R. D., Ericksen, S., Chang, S., & Gallagher, P. (1985). Efficacy of ECT: A meta-analysis. *American Journal of Psychiatry, 142,* 297-302.

Vlahov, D., Anthony, J. C., Celentano, D., Solomon, L., & Chowdhury, N. (1991). Trends of HIV-1 risk reduction among initiates into intravenous drug use 1982-1987. *American Journal of Drug and Alcohol Abuse, 17,* 39-48.

Wright, L. J., & Carlquist, J. F. (1994). Measles immunity in employees of a multihospital health care provider. *Infection Control and Hospital Epidemiology, 15,* 8-11.

The Median Test

The median test is used when the independent variable is categorical, the dependent variable is at least at the ordinal level of measurement, and the researcher wants to examine differences between two or more groups with regard to their measures of central tendency, specifically the median (Conover, 1980). It answers the question as to whether two or more groups come from populations with identical distributions. The median test is particularly appropriate when the exact values of extreme scores have been truncated to above or below a certain cutoff point (Siegel & Castellan,

205

1988). It also can be used when the assumption of similarity of shapes of distributions of the Mann-Whitney U test (Chapter 7) has not been met.

In our hypothetical study, the categorical dependent variable, post-hospital adjustment, had been collapsed from a 100-point scale into three categories: poor, moderate, and excellent posthospital adjustment. Using these collapsed data, both the ordinary chi-square test and the Mantel-Haenszel chi-square test for trends were used to examine the relationship between social status and adjustment. A problem with collapsing data is that important information potentially can be lost in the categorizations. For example, the values 70 and 71 on this test are within one point of each other and are closer to each other than are the values 55 and 80, yet a child who scored 70 on the test was placed in the same "poor" adjustment category as the child whose score was 55, whereas the child with a score of 71 was ranked as having "moderate" adjustment similar to that of a child with a score of 80.

Suppose that we elected not to collapse the data but to retain the variable "posthospital adjustment" as an interval-level variable and to compare these data across the three social status groups. Now we have an independent variable, social status, that is at the nominal level and a dependent variable, posthospital adjustment, that is continuous. Because the dependent variable is continuous, we can no longer use chi-square tests to evaluate the relationship between family social status and posthospital adjustment. We use instead either the median test or the Kruskal-Wallis one-way ANOVA.

**An Appropriate Research Question
for the Median Test**

The median test has not been a popular nonparametric statistic in recent years. This may be due, in part, to the advent of personal computers with user-friendly statistical packages that have facilitated the calculation of more powerful nonparametric statistics, including the Mann-Whitney test and the Kruskal-Wallis one-way ANOVA by ranks. There are, however, a few recent studies that have made use of this handy statistic. For example, Hojat, Gonnella, and Xu (1995) used the median test to compare 667 men and women on their evaluations of their medical education, professional life, and clinical practice. Allen, Weisman, Weedon, and Krasinski (1993) used the same test to examine factors that might affect perinatal HIV transmission in 70 HIV-positive women, and Jeffery, Wardlaw, Nelson Collins, and Kay (1989) used the median test to assess structural changes in bronchial biopsies from 11 asthmatic patients and 10 control subjects.

Table 8.7 Example of Null and Alternative Hypotheses Suitable for Use With the Median and Kruskal-Wallis Tests

---

Null Hypothesis

H_0:   There are no differences in medians among the three family social status groups (low, medium, or high) with regard to a mother's evaluation of her child's post-hospital adjustment; that is, all three groups have similar underlying distributions of posthospital adjustment scores.

Alternative Hypothesis

H_a: There are differences in medians among the three family social status groups (low, medium, or high) with regard to a mother's evaluation of her child's posthospital adjustment; that is, at least one pair of groups differs with respect to its underlying distribution of posthospital adjustment scores.

In our hypothetical study, we could use the median test to answer a number of different research questions. A typical example would be:

Are there differences among the three social status groups (low, medium, and high social status) with regard to mothers' evaluations of their children's posthospital adjustment?

Null and Alternative Hypotheses

Table 8.7 presents examples of possible null and alternative hypotheses derived from our research question that would be suitable for the median test. Note that the alternative hypothesis is nondirectional, making this a two-tailed test. It is also possible to predict the direction of the medians. This would necessitate using either a one-tailed median test, if only two groups are involved, or a priori comparisons, if there are more than two groups.

Overview of the Procedure

To undertake the median test, the data are first considered as a single sample, and the combined *grand* median for the scores on the dependent variable is calculated. Next, the sample is separated into the various levels of the independent variable. If the null hypothesis is true and the samples share the same median, then we would expect that about half the observations within each level of the independent variable would be above and below the *grand median*.

The next step, therefore, is to classify the observations within each group according to whether observations lie above or below the grand median. These frequency counts are then arranged in a 2 × c contingency table where the 2 rows represent the 2 categories, *greater than* or *less than or equal to* the median, and the c columns are the levels of the independent variable.

Finally, the observed number of observations above and equal to/below the median for each of the levels of the independent variable, O_{ij}, are compared to what would have been expected under the null hypothesis of equal medians. These values are then summed across all cells. The test statistic used to evaluate this 2 × c contingency table is the chi-square statistic:

$$\chi^2 = \sum_{i=1}^{2} \sum_{j=1}^{c} \frac{(O_{ij} - E_{ij})^2}{E_{ij}}$$

where

O_{ij} = the observed number of observations in a cell in row i and column j

E_{ij} = the expected number of observations in the same cell if the medians were equal.

If the data meet the assumptions of the median test, this chi-square statistic is distributed as a χ^2 with $(c - 1)$ degrees of freedom.

The decision rule is to reject the null hypothesis of equal medians if this calculated χ^2 value exceeds the critical χ^2 value at $df = (c - 1)$ and the prestated α level. Alternatively, when presented with computer printouts, the decision rule would be to reject the null hypothesis if the generated significance level (p) is less than the predetermined α level.

Critical Assumptions of the Median Test

There are few assumptions associated with the median test, which makes it a user-friendly nonparametric test. Other than the fact that the data have been obtained from a random sample, the following assumptions apply:

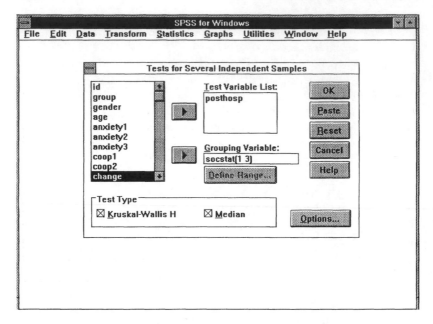

Figure 8.3. SPSS for Windows Dialogue Box for the Median and Kruskal-Wallis Tests

1. The independent variable is at the nominal level of measurement. The dependent variable is at least at the ordinal level of measurement.

2. The levels of the independent variable are independent of one another; that is, a respondent can belong to one and only one sample of observations.

Our data from the hypothetical study meet all the assumptions except for its being a random sample. The independent variable, social status, is at the nominal level of measurement with three mutually exclusive levels, and the dependent variable, posthospital adjustment, is at the interval level of measurement.

Computer Commands

Figure 8.3 presents the SPSS for Windows dialogue box that is used to undertake the median test. This box was opened by choosing the following items from the menu: *Statistics . . . Nonparametric Tests . . . k Independent Samples.* Note that the minimum and maximum values for the independent variable (social status) need to be specified through the *Define Range*

Table 8.8 Computer-Generated Printout Obtained From SPSS for
Windows for the Median Test

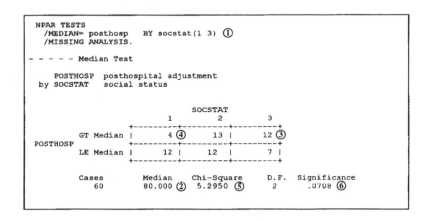

command. In our hypothetical study, these values are 1 (low social status)
and 3 (high social status).

Computer-Generated Output

The syntax commands (①) and computer-generated output for the median test are presented in Table 8.8. The contingency table that is presented outlines, for each group, the number of subjects with posthospital adjustment values greater than or equal to/less than the grand median of 80.00 (②). Table 8.8 indicates a discrepancy of distribution of scores above and below the median for both the lower and higher status groups and a somewhat equal distribution of scores for the middle status group. For example, 12 of 19 children in the high social status group received posthospital adjustment scores that were greater than the median (③) compared with only 4 of 16 children in the lower status group (④). Are the discrepancies sufficiently large to reject the null hypothesis of similarity of medians?

Following our decision rule, we will reject the null hypothesis of equal medians if our actual χ^2 exceeds a critical χ^2 at $df = 2$ ($df = c - 1 = 3 - 1 = 2$) with $\alpha = .05$. From tables that provide the critical values of the chi-square

Table 8.9 Presentation of the Results of the Median Test Analyses of
Posthospital Adjustment in Table Form

					Median Test	
Social Status	N	Mean	Median	Range	χ^2	p
Low	16	75.8	77	65 – 89		
Middle	25	78.5	81	65 – 85	5.29	.07
High	19	82.6	85	62 – 96		

distribution (e.g., Table C, Siegel & Castellan, 1988), we can see that this $\chi^2_{cv} = 5.99$. Alternatively, if we are examining the computer printout, we will reject the null hypothesis of equal medians if the generated p value or significance level is less than .05.

From Table 8.8, we can see that the generated χ^2 value is 5.295 (⑤), which is less than our $\chi^2_{cv} = 5.99$. Similarly, our p value, .07 (⑥), is greater than our prestated alpha (.05); therefore, we cannot reject the null hypothesis of similarity of medians. The conclusion to be drawn from the median test, therefore, is that there are no differences among the three groups with regard to their median posthospital adjustment scores.

Recall that the median test, like the one-way ANOVA and the chi-square test, is an overall test of significance. The median test does not indicate where the differences among the groups lie. Had we been able to reject the null hypothesis, we would want to undertake post hoc tests to determine which of the group medians were significantly different from one another. Examples of post hoc tests are presented with the Kruskal-Wallis test in this chapter.

Presentation of Results

Table 8.9 is an example of how the results of the median test could be presented. To better describe the descriptive characteristics of the sample, means, medians, and ranges are presented. The table also indicates that the presented χ^2 value is the result of the median test in order to prevent confusion with other chi-square statistics. The results of this test could also be interpreted and presented in the text:

The results of the median test indicate that there were no significant differences among the three social status groups with regard to their median posthospital adjustment scores ($p = .07$).

Advantages and Limitations of the Median Test

The median test has the advantage of being very straightforward and easy to apply, a characteristic that was especially welcomed during the precomputer era. It is a particularly useful test when the researcher does not know the exact values of all the scores, especially those at the extremes. The limitation of this test is that it considers only two possibilities for scores: They are either above or below/equal to the median. The size of the differences between the observed scores and the median is not taken into account. This results in the median test being a less powerful test than tests that do consider the size of differences, such as the Mann-Whitney test (Chapter 7) and the Kruskal-Wallis one-way ANOVA by ranks.

Parametric and Nonparametric Alternatives to the Median Test

There are two parametric alternatives to the median test: the t test when the independent variable is dichotomous and the one-way ANOVA when the independent variable has more than two levels. The nonparametric alternatives to the median test are the Mann-Whitney test for two groups and the Kruskal-Wallis one-way ANOVA for more than two groups. When the exact range of values for the dependent variable is known, the Mann-Whitney and Kruskal-Wallis tests are the preferred nonparametric tests because they take into account the size of the differences between the observed scores and the grand median.

Examples From Published Research

Allen, M., Weisman, C., Weedon, J., & Krasinski, K. (1993). Antepartum and intrapartum factors in perinatal HIV transmission. *International Conference on AIDS, 9,* 649.

Hojat, M., Gonnella, J. S., & Xu, G. (1995). Gender comparisons of young physicians' perceptions of their medical education, professional life, and practice: A follow-up study of Jerrerson Medical College graduates. *Academic Medicine, 70,* 305-312.

Jeffery, P. K., Wardlaw, A. J., Nelson, F. C., Collins, J. V., & Kay, A. B. (1989). Bronchial biopsies in asthma: An ultrastructural, quantitative study and correlation with hyperreactivity. *American Review of Respiratory Disease, 140*, 1745-1753.

The Kruskal-Wallis
One-Way ANOVA by Ranks

Like the median test, the Kruskal-Wallis (K-W) one-way ANOVA by ranks is a test that is used to determine whether k independent samples come from a population with a common median. As an extension of the median test, the K-W test is used when the independent variable is at the categorical level of measurement, with more than two levels, and the dependent variable is at least ordinal. It is more powerful than the median test, however, because it makes more complete use of the information contained in the observations. The K-W test ranks the values instead of merely noting them as being above or below the median. It is interesting that the K-W test was derived from the one-way ANOVA, with the actual observations being replaced by their ranks (Kruskal & Wallis, 1952).

An Appropriate Research Question
for the Kruskal-Wallis Test

Numerous studies in the health care field have used the K-W test. For example, Mange and colleagues (1992) used the K-W test to compare recovery rates of 39 spinal cord injured patients. Watkins and Kligman (1993) used a K-W one-way ANOVA by ranks to compare attendance patterns of 224 older adults who took part in a health promotion program. Packer, Sauriol, and Brouwer (1994) used it to examine the severity of fatigue and its impact on patients with postpolio syndrome, chronic fatigue syndrome, and multiple sclerosis, and Richmond, Metcalf, and Daly (1992) used the same test to determine nursing care costs in the acute phase of spinal cord injury. Livingston, Healy, Jordan, Warner, and Zazzali (1994) used the K-W test to determine the perceived needs of 790 perimenopausal women and their clinicians regarding management of menopause.

In our hypothetical study, the research question that was addressed using the median test could also be examined using the K-W one-way ANOVA by ranks:

What are the differences between the three social status groups (low, medium, and high social status) with regard to mothers' evaluations of their children's posthospital adjustment?

Although other questions could have been answered using this test, we are duplicating the analysis here to demonstrate the added power of the Kruskal-Wallis test. This will also enable us to compare these results with those we obtained when we analyzed the collapsed posthospital adjustment data using the chi-square and Mantel-Haenszel tests.

Null and Alternative Hypotheses

Because we are interested in comparing the results obtained from the median and K-W tests, the null and alternative hypotheses for both tests are presented in Table 8.7. The null hypothesis states that the three social status groups come from a population with the same median, whereas the alternative hypothesis states that at least one of the groups comes from a population with a different median.

Overview of the Procedure

To calculate a K-W test, the data for the entire sample are first ranked from lowest to highest. The smallest score in the sample is given a rank of "1," and the highest score is given the rank of N, where N represents the total sample size. Next, the scores in the separate cells are replaced by their rankings and the average sum of the ranks for each cell is calculated. If the null hypothesis that the independent groups have similar underlying population distributions is true, the average sum of the ranks for each group should be the same. If the discrepancy between the average ranks for the independent groups is sufficiently large, the null hypothesis will be rejected.

The K-W statistic that will be used compares the average rank for the separate cells to the average rank for the entire sample using the following formula (Siegel & Castellan, 1988):

$$K - W = \frac{12}{N(N+1)} \sum_{j=1}^{k} n_j (\overline{R}_j - \overline{R})^2$$

where

K = the number of groups

n_j = the number of subjects in the jth group

N = the total sample size

$\overline{R_j}$ = the average of the ranks for the jth group

$\overline{R}$ = $(N + 1)/2$ = the average of the ranks for the entire sample.

If the data meet the assumptions of the Kruskal-Wallis one-way ANOVA by ranks and the sample size is sufficiently large (greater than 5 subjects per group), this K-W statistic is approximately distributed as a χ^2 with $df = (k - 1)$. The null hypothesis of equality of groups will be rejected if the generated K-W statistic is greater than the critical value of the χ^2 at a prestated level of alpha (e.g., $\alpha = .05$) with $df = (k - 1)$. Alternatively, if presented with a computer printout, the null hypothesis will be rejected if the observed p value is less than the previously set α level.

Critical Assumptions of the Kruskal-Wallis Test

The Kruskal-Wallis test shares many of the assumptions of the Mann-Whitney test (Chapter 7) but extends these assumptions to more than two independent groups.

1. *The data have been collected from a randomly selected set of observations.*
2. *The dependent variable is at least at the ordinal level of measurement.*
3. *The independent variable is nominal, with more than two levels.*
4. *There is independence of observations within each group and between groups. There are no repeated measures or multiple response categories.*
5. *The shapes of the distributions of the dependent variable within each of the groups are similar except for a possible difference in measure of central tendency of at least one of the groups.*

The data from our hypothetical study partially meet these assumptions in that the dependent variable is at the interval level of measurement, the independent variable is categorical with three levels, and there are no repeated measures or multiple response categories. The similarity in shapes of the distributions for the three social status groups was determined by examining the histograms generated in the *Explore* command. Although they were rather skewed, they did have similarity in shapes. This is not a random sample, however.

Computer Commands

The median and Kruskal-Wallis tests are generated from the same dialogue box in SPSS for Windows (Figure 8.3). This dialogue box is opened by choosing the following items from the menu: *Statistics*. *Analyze Nonparametric Tests . . . K Independent Samples*, inserting the dependent variable (*test variable*) and independent variable (*grouping variable*), and indicating the minimum and maximum values for the grouping variable (e.g., "1" and "3") using the *Define Range* dialogue box. Note that if the researcher were interested in comparing only Groups 4, 5, and 6 in a six-level categorical variable, the minimum and maximum values of 4 and 6 would be specified.

Computer-Generated Output

Table 8.10 presents the syntax commands ① and computer-generated output for the K-W one-way ANOVA by ranks. The mean ranks for the three social status groups ② indicate that the children in the low social status group have the lowest mean rank for posthospital adjustment, whereas the children in the high social status group have the highest rank.

Except for slight round-off error, the K-W calculated χ^2 value presented in the printout (6.3689, ③) is similar to that obtained using the formula for the K-W statistic:

$$K - W = \frac{12}{N(N+1)} \sum_{j=1}^{k} n_j (\bar{R}_j - \bar{R})^2$$

$$= \frac{12}{60(60+1)} [16 (23.07-30.5)^2 + (25)(29.65-30.5)^2 + (19)(37.88-30.5)^2]$$

$$= .0033 [883.27 + 18.06 + 1034.82]$$

$$= 6.38$$

Table 8.10 Computer-Generated Printout for the Kruskal-Wallis One-Way ANOVA by Ranks

```
NPAR TESTS
 /K-W=posthosp    BY socstat(1 3)   ①
 /MISSING ANALYSIS.

- - - - - Kruskal-Wallis 1-Way Anova

   POSTHOSP   posthospital adjustment
 by SOCSTAT   social status


   Mean Rank    Cases

② 23.07         16   SOCSTAT =    1   low
   29.65         25   SOCSTAT =    2   middle
   37.88         19   SOCSTAT =    3   high

                 --

                 60   Total

                                        Corrected for ties
   Chi-Square      D.F.  Significance   Chi-Square   D.F.  Significance
     6.3689 ③        2      .0414 ⑤       6.3956 ④    2       .0409 ⑤
```

In the printout, we are also presented with an adjusted χ^2 value that is corrected for the influence of tied observations (6.3956, ④). Correcting for ties increases the value of the χ^2, which in turn enhances the likelihood that the null hypothesis will be rejected. Siegel and Castellan (1988) indicate that if fewer than 25% of the observations are tied, the value of the K-W statistic will not be changed substantially. This appears to be the case in our example (Table 8.10): The χ^2 value did not change dramatically.

To determine whether the K-W statistic is sufficiently large to reject the null hypothesis of equal medians, the obtained value of the χ^2 is compared to a critical χ^2 value with $df = 2$ ($k - 1 = 3 - 1 = 2$). If the obtained value of the χ^2 is larger than the critical value, the null hypothesis of equal medians will be rejected. Alternatively, using the printout, the null hypothesis will be rejected if the obtained p value (significance) is *less than* the prestated alpha (e.g., $\alpha = .05$).

The critical value of a χ^2 with $df = 2$ and $\alpha = .05$ is 5.99 (Table C, in Siegel & Castellan, 1988). This value is smaller than our obtained value for both the uncorrected and corrected χ^2 values (6.3689 and 6.3956, respectively). Similarly, the obtained p values for both χ^2 values (.0414 and .0409) are less than the prestated $\alpha = .05$ (⑤). We will, therefore, reject the null hypothesis of similarity of medians and conclude that at least one of the groups is significantly different from the others.

Determining Which Groups Are Significantly Different

Like the one-way ANOVA, the Kruskal-Wallis test is an omnibus test of significance in that, given a significant result, the test does not indicate where the differences are among the groups. To determine which groups are significantly different from one another, it is necessary to undertake post hoc comparisons. Two approaches to post hoc comparisons will be presented here: the Dunn multiple comparisons procedure and pairwise comparisons using the Mann-Whitney test.

The Dunn Multiple-Comparisons Procedure

The Dunn procedure (Dunn, 1964; Neave & Worthington, 1988) is a very effective, though somewhat conservative, post hoc approach to testing pairwise comparisons among groups. Unfortunately, SPSS for Windows (V6.1) does not offer this procedure, although it is easily performed using a hand calculator.

Once a determination has been made that there is at least one significant difference among the groups being tested (i.e., the Kruskal-Wallis test is significant), the differences in the mean ranks between the groups are computed pairwise. Next, a standard deviation (σ_i) for each of these comparisons is calculated. From these values, a t statistic for each pairwise comparison is generated by dividing the absolute difference in ranks between each of the pairs by its standard deviation. This t statistic, which is distributed approximately normally, is compared to the standard normal z distribution. If the resulting t statistic at $\alpha/2$ is sufficiently large, the null hypothesis of no differences between the ranks is rejected. This t statistic is calculated as follows:

$$T_{ij} = \frac{|D_{ij}|}{\sigma_{if}} = \frac{|\bar{R}_i - \bar{R}_j|}{\sigma_{if}} = \frac{|\bar{R}_i - \bar{R}_j|}{\sqrt{\frac{N(N+1)}{12}\left[\frac{1}{n} + \frac{1}{n_j}\right]}}$$

Table 8.11 Calculations for the Dunn Multiple Comparison Procedure

Group	Mean Rank Difference $\|D_{ij}\|$	Standard Deviation of Differences $\sigma_{ij} = \sqrt{\dfrac{N(N+1)}{12}\left[\dfrac{1}{n_i}+\dfrac{1}{n_j}\right]}$	t Statistic $\|t_{ij}\|$ ①
1 vs. 2	$\|23.07 - 29.65\| = 6.58$	$\sqrt{\dfrac{60(60+1)}{12}\left[\dfrac{1}{16}+\dfrac{1}{25}\right]} = 5.59$	$6.58/5.59 = 1.18$
1 vs. 3	$\|23.07 - 37.88\| = 14.81$	$\sqrt{\dfrac{60(60+1)}{12}\left[\dfrac{1}{16}+\dfrac{1}{19}\right]} = 5.93$	$14.81/5.93 = 2.50$
2 vs. 3	$\|29.65 - 37.88\| = 8.23$	$\sqrt{\dfrac{60(60+1)}{12}\left[\dfrac{1}{25}+\dfrac{1}{19}\right]} = 5.31$	$8.23/5.31 = 1.55$

where

$\|\overline{R}_i - \overline{R}_j\|$ = the absolute value of the difference in mean ranks between Group i and Group j

N = the total sample size

n_i = the sample size for Group i

n_j = the sample size for Group j

In our hypothetical example, there are three pairwise comparisons to be made: Group 1 versus Group 2, Group 1 versus Group 3, and Group 2 versus Group 3. Table 8.11 presents the calculations and t values for each of these comparisons.

Table 8.11 indicates that the t values for the three comparisons are 1.18, 2.50, and 1.55, respectively (①). To determine which pairs, if any, are significantly different from one another, it is necessary to compare the obtained t value for each of the three comparisons with the absolute value of a critical z at a predetermined level of alpha ($\alpha/2$ because this is a two-tailed test and we are dealing with absolute values). The null hypothesis of similarity of mean ranks for the pair of comparisons will be rejected if the actual t value is greater than the critical z at $\alpha/2$.

What α should be used to examine these pairwise comparisons? Because there are multiple comparisons being made on the same sample, we need to adjust our α to accommodate a potentially inflated Type I error. We could adjust the α level using Bonferroni's inequality where $\alpha' = \alpha/k$ (k = the number of tests) or use a slightly more liberal stepdown approach suggested by Holms (Holland & Copenhaver, 1988; Stevens, 1990). To illustrate the differences between the two approaches, Bonferroni's inequality will be used for the Dunn procedure and the Holms procedure will be used for the Mann-Whitney post hoc comparisons.

If our original alpha were $\alpha = .05$, then $\alpha' = \alpha/k = .05/3 = .016$, which is very restrictive. Dunn (1964) and Neave and Worthington (1988) argue that, when undertaking post hoc comparisons, α should be more liberal than the usual $\alpha = .05$ (e.g., .10 or .20) so that any differences between groups can be detected. These authors point out that multiple comparisons should be made only once a significant overall test has been obtained at the usual α (e.g., $\alpha = .05$). This gives a good protection against inflated error when using a more liberal α for the post hoc tests.

Suppose we decided that, given the significant overall test at $\alpha = .05$, we would increase our α to .10 for the post hoc tests to better detect differences among the groups. Using Bonferroni's inequality, we would then undertake each post hoc test at $\alpha' = .10/3 = .033$. The critical value of the z statistic, therefore, for $\alpha'/2 = .017$ is 2.128, because we are undertaking a two-tailed test. The null hypothesis of similarity of mean ranks will be rejected, therefore, for any pair of comparisons in which $t_{ij} \geq 2.128$.

Table 8.11 indicates that there is only one comparison that meets this criteria: Group 1 versus Group 3, the low versus high status groups (②). Our conclusion using the Dunn procedure, therefore, is that the children in the high social status group were rated by their mothers to be have significantly higher posthospital adjustment than the children in the low social status group.

Pairwise Comparisons Using the Mann-Whitney Test

An alternative test for post hoc comparisons is the Mann-Whitney test. This test has the advantage of being readily available in statistical packages (for details of the Mann-Whitney test, see Chapter 7) and can be obtained in SPSS for Windows by selecting the following items from the menu: *Statistics . . . Nonparametric Tests . . . 2 Independent Samples*. The particular groups to be compared (e.g., 1 vs. 2, 1 vs. 3, etc.) can be specified by highlighting the independent (grouping) variable, clicking on the *Define Range* dialogue box, and selecting the specific groups to be compared.

Table 8.12 Post Hoc Comparisons of the Three Social Status Groups
Using the Mann-Whitney Test

```
- - - Mann-Whitney U - Wilcoxon Rank Sum W Test

  POSTHOSP  posthospital adjustment
by SOCSTAT  social status

  Mean Rank    Cases

     13.81      16  SOCSTAT = 1.00  low
     21.53      19  SOCSTAT = 3.00  high
                --
                35  Total
                            Exact             Corrected for ties
       U            W      2-Tailed P         Z       2-Tailed P
      85.0        221.0     .0243           -2.2466     .0240  ②

*************************************************************************
- - - - - Mann-Whitney U - Wilcoxon Rank Sum W Test

  POSTHOSP  posthospital adjustment
by SOCSTAT  social status

  Mean Rank    Cases

     17.75      16  SOCSTAT = 1.00  low
     23.08      25  SOCSTAT = 2.00  middle
                --
                41  Total
                     ①
                            Exact             Corrected for ties
       U            W      2-Tailed P         Z       2-Tailed P
     148.0        284.0     .1706           -1.3937     .1634

*************************************************************************
- - - - - Mann-Whitney U - Wilcoxon Rank Sum W Test

  POSTHOSP  posthospital adjustment
by SOCSTAT  social status

  Mean Rank    Cases

     19.56      25  SOCSTAT = 2.00  middle
     26.37      19  SOCSTAT = 3.00  high
                --
                44  Total
                             Corrected for ties
       U            W         Z       2-Tailed P
     164.0        501.0     -1.7466     .0807  ③
```

These results and their significance levels (*p* values) for our hypothetical
example are presented in Table 8.12. Note that, for two of the three tests,
two significance levels are presented: an exact two-tailed *p* value and a *p*
value corrected for ties. Norusis (1995a) indicates that this exact *p* value
is presented when the total sample size is small (e.g., *N* < 30). It is
interesting to note that, for our hypothetical example, this exact *p* value is

presented when $N = 41$ (①), but not for $N = 44$. Because our sample is reasonably large ($N > 35$ for each comparison) and we undoubtedly have tied values, our interest will be on the p value for the z statistic that corrects for these ties. Looking at these p values, it appears that if we were to use a more liberal two-tailed $\alpha = .10$, two significant differences arise: the group differences between low and high social status (②) and between middle and high status (③). The only problem with these Mann-Whitney tests is that we have not adjusted for the inflated Type I error created as a result of undertaking three tests on the same set of data.

The Holms Stepdown Procedure for Post Hoc Comparisons

The Holms stepdown procedure will be used here to illustrate a second approach to adjusting for inflated Type I error. Using this procedure, the generated p values for the Mann-Whitney post hoc tests are first ranked from lowest to highest ($p_1 < p_2 < p_3 < \ldots p_i$). These p values are then compared to $\alpha/(k - i + 1)$, where α = the prestated alpha level (e.g., .10), k = the number of post hoc tests generated, and i = the index of the particular p value that is being compared. The null hypothesis for a particular test will be rejected if $p_i < \alpha/(k - i + 1)$.

Table 8.13 presents the results of the Holms procedure for the three post hoc tests in our hypothetical study. Note that, following the Holms procedure, we have switched the order of the last two tests given in Table 8.12 because the p value for the medium versus high groups (.081) is smaller than that for the low versus medium groups (.163). These results indicate that the only significant difference among the groups is between the low and high social status groups (①) because .024 < .033. This result suggests that the children in the high social status group received higher posthospital adjustment scores than did the children in the lower social status group. The significance values for the other comparisons were both greater than our adjusted alpha levels, indicating that no other groups were significantly different. This is the same conclusion we reached using the Dunn multiple-comparisons procedure.

Presentation of Results

The results of the Kruskal-Wallis test could be presented in a table similar to that used for the median test (Table 8.9). Instead of the χ^2 value for the median test, however, the K-W χ^2 value is presented with its corresponding p value (e.g., 6.37, $p = .04$). The results, along with the post hoc comparisons, also could be presented in the text as follows:

Table 8.13 Results of the Holms Stepdown Procedure to Adjust for Type I
Error When Undertaking Multiple Comparisons

i	Group	Comparison	Obtained p value[a]	$\alpha/(k - i + 1)$
1	1 vs. 3	low vs. high	.024 ①	$.10/(3 - 1 + 1) = .033$
2	1 vs. 2	middle vs. high	.081	$.10/(3 - 2 + 1) = .05$
3	2 vs. 3	low vs. middle	.163	$.10/(3 - 3 + 1) = .10$

a. Mann-Whitney tests, Table 8.12.

The results of the Kruskal-Wallis test indicate that there were significant differences among the three social status groups with regard to mothers' evaluations of their children's posthospital adjustment ($\chi^2_{K-W} = 6.37, p = .04$). Post hoc analyses using Mann-Whitney tests indicated that children of lower social status were evaluated by their mothers as having poorer posthospital adjustment than were children of higher social status ($p = .024$). No other groups had significantly different median adjustment scores.

Advantages and Limitations of the Kruskal-Wallis Test

The Kruskal-Wallis one-way ANOVA by ranks is a simple test to understand and use. Like the one-way ANOVA, the K-W test does not require equal sample sizes. It is also one of the most popular and powerful of the nonparametric tests for continuous data.

Parametric and Nonparametric Alternatives to the Kruskal-Wallis Test

The parametric alternative to the Kruskal-Wallis test is the one-way ANOVA. This test assumes normality of distributions within each level of the independent variable and equality of variances among levels of the independent variable. The ANOVA is more powerful than the K-W test when its somewhat stringent assumptions have been met. Although the ANOVA is robust with regard to normality of distributions, it is less reliable when the assumption of equal variances has been violated. Moreover, the K-W test has power that is almost equal to that of the ANOVA for normal distributions and is more powerful than the parametric statistic when the distribution of the data is skewed (Dexter, 1994; Gibbons, 1985; Neave & Worthington, 1988). Daniel (1990) reports that the asymptotic efficiency of the K-W test relative to the *F* test for the one-way ANOVA is .955 if the distribution of the dependent variable within the groups is normal and not less than .864 when the shapes of the distributions are similar.

Alternative nonparametric tests that could be used in place of the K-W test are those that were reviewed in this chapter: the median test, the Mantel-Haenszel chi-square test, and the chi-square test for *k* independent samples. In the summary that follows, the relative power of these three tests will be compared.

Examples From Published Research

Livingston, W. W., Healy, J. M., Jordan, H. S., Warner, C. K., & Zazzali, J. L (1994). Assessing the needs of women and clinicians for the management of menopause in an HMO. *Journal of General Internal Medicine, 9,* 385-389.

Mange, K. C., Marino, R. J., Gregory, P. C., Herbison, G. J., & Ditunno, J. F. (1992). Course of motor recovery in the zone of partial preservation in spinal cord injury. *Archives of Physical Medicine and Rehabilitation, 73,* 437-441.

Packer, T. L., Sauriol, A., & Brouwer, B. (1994). Fatigue secondary to chronic illness: Postpolio syndrome, chronic fatigue syndrome, and multiple sclerosis. *Archives of Physical Medicine and Rehabilitation, 75,* 1122-1126.

Richmond, T., Metcalf, J., & Daly, M. (1992). Cost of nursing care in acute spinal cord injury (SCI). *Heart and Lung: Journal of Critical Care Abstract, 21,* 293.

Watkins, A. J., & Kligman, E. W. (1993). Attendance patterns of older adults in a health promotion program. *Public Health Reports, 108,* 86-90.

Summary

In this chapter, we examined four tests to be used when the independent variable had more than two groups: the chi-square test for *k* independent samples, the Mantel-Haenszel chi-square test for ordered categories, the median test, and the Kruskal-Wallis test. The first two of these tests are intended to be used when the dependent variable is categorical. The last two tests are to be used when the dependent variable is continuous.

To compare the results generated by these four tests, we used the same independent variable (social status). The dependent variable (children's posthospital adjustment) was collapsed into three groups (poor, moderate, and excellent posthospital adjustment) for the first two tests but was retained as a continuous variable for the last two tests. Table 8.14 summarizes the results for these four tests.

Table 8.14 Comparison of the Power of the Four Tests Reviewed in This
Chapter

Test	Independent Variable	Dependent Variable	p Value	Decision
Chi-square	Nominal	Nominal	.049	Reject H_0
Mantel-Haenszel chi-square	Nominal	Nominal	.043	Reject H_0
Median	Nominal	Continuous	.071	Fail to reject H_0
Kruskal-Wallis	Nominal	Continuous	.041	Reject H_0

As can be seen from Table 8.14, the null hypothesis was rejected in all but the median test. The K-W test has the smallest resulting p value, suggesting that it was the most sensitive test.

The median test and the K-W test have the same asymptotic χ^2 distribution; that is, if the sample size is sufficiently large, both statistics are approximately distributed as a χ^2 with $df = k - 1$. The more powerful of these two tests, therefore, will be that test which generates the larger χ^2 value. In Table 8.8, we saw that the χ^2 value for the median test was 5.29. In Table 8.10, the χ^2 value for the K-W test is larger (6.3689). Based on the two test results and their p values (Table 8.14), we made different decisions regarding the null hypothesis. With the median test, we concluded that there were no differences among the group medians and failed to reject the null hypothesis. Using the K-W test, we concluded that there were differences among the groups and rejected the null hypothesis. These seemingly conflicting results support the contention that the Kruskal-Wallis test may be more sensitive than the median test in that it may be more likely to correctly reject the null hypothesis.

9 Tests of Association Between Variables

- **Phi coefficient**
- **Cramér's *V* coefficient**
- **Kappa coefficient**
- **Point bisenial correlation**
- **Spearman rho correlation**
- **Kendall's tau**

It frequently occurs in health care research that we are interested in measuring the degree of association or correlation between two variables. In our hypothetical study, for example, we might want to examine the relationship between a child's age and his or her level of posthospital adjustment. Depending on the level of measurement of the two variables being examined, there are a number of nonparametric tests that can provide information about the extent of the relationship between variables. In this chapter, we will examine six bivariate measures of association: the phi, Cramér, and kappa coefficients for two categorical variables; the point biserial correlation for examining the relationship between a dichotomous and a continuous variable; and the Spearman rho rank-order correlation and Kendall's tau coefficients for two continuous variables.

The Phi Coefficient

It was pointed out in Chapter 8 that the phi coefficient is a useful statistic for determining the strength of relationship between two dichotomous variables after a chi-square test of association or Fisher exact test for small samples has produced a significant result. In this section, we will examine this coefficient in greater depth.

An Appropriate Research Question for the Phi Coefficient

The phi coefficient serves a number of useful functions. It can be used when the researcher is interested in testing hypotheses about the degree and strength of the relationship between two variables that are dichotomous, such as gender (male, female) and test results (pass, fail). It is typically used after a significant result has been obtained from the chi-square test for two independent samples. It also has been used as a measure of reliability for two-category nominal scales (Conger & Ward, 1984) and to evaluate the suitability of multiple-choice questions in examinations (Koeslag, Schach, & Melzer, 1987).

A number of examples in the health care literature demonstrate the versatility of the phi coefficient. Krettek, Arkin, Chaisilwattana, and Monif (1993) used the phi coefficient to identify the incidence of positive assays indicating *Chlamydia trachomatis* in oral contraceptive users with ($n = 65$) and without ($n = 65$) spotting and in women seeking contraception ($n = 65$). Stolovitzky and Todd (1990) used the same statistic to evaluate the incidence of otitis media in short-headed and long-headed persons. Bond and Monson (1984) used the phi coefficient to examine the effectiveness of an intervention program involving a clinical pharmacist and nurse clinician in improving drug documentation, patient compliance, and disease control in an ambulatory care population. In a very different context, Eaves and Milner (1993) used this same statistic to assess the criterion-related validity of two autism rating scales.

The hypothetical example from Chapter 7 will be used to illustrate the approach to evaluating the strength of the phi coefficient when the two variables of interest are at the nominal level of measurement. In Chapter 7 (Table 7.7), the chi-square test indicated that the two variables, group membership (intervention, control) and children's expressed postintervention fear of their impending operations (yes, no) were not independent ($\chi^2 = 4.82$, p = .028). We concluded, therefore, that the two variables were associated. Because the chi-square test for independence does not evaluate the strength of that association, we used the phi coefficient to determine this. A research question that could be answered using the phi coefficient, therefore, would be:

What is the strength of association between the variables of group membership (intervention, control) and children's expressed fear of their impending operations (yes, no)?

Table 9.1 Example of Null and Alternative Hypotheses Suitable for Testing With a Phi Coefficient

---

Null Hypothesis

H_0:   There is no association between the variables of group membership (intervention, control) and children's expressed fear of their impending operations (yes, no).

Alternative Hypothesis

H_a: There is an association between the variables of group membership (intervention, control) and children's expressed fear of their impending operations (yes, no).

Null and Alternative Hypotheses

Table 9.1 presents examples of null and alternative hypotheses generated from the research question outlined above that could be analyzed using the phi coefficient. Note that, as for the chi-square statistic, the null and alternative hypotheses for the phi coefficient reflect the association between two dichotomous variables.

Overview of the Procedure

To calculate the phi coefficient, the variables of interest are first arranged in a 2 × 2 table in which the data represent frequencies, not scores (Table 9.2). The variables, X and Y, represent the independent and dependent variables whose categories have been assigned the values of 0 and 1, and the values of a through d represent the frequencies for these categories.

From this 2 × 2 contingency table, a χ^2 statistic (Chapter 7) is calculated. The phi coefficient (ϕ) is obtained either directly from the 2 × 2 table or by taking the square root of the χ^2 value divided by the total sample size:

$$f = \frac{(ad - bc)}{\sqrt{(a+b)(c+d)(a+c)(b+d)}} = \sqrt{\chi^2/N}$$

Because ϕ is based on a χ^2 value, the significance level for this coefficient is the same as for the χ^2 statistic. The null hypothesis of no association will be rejected if the calculated χ^2 value is greater than the critical χ^2 at $df = 1$ or, alternatively, if the generated p value is less than the prestated α level (e.g., .05).

Table 9.2 Example of a 2 × 2 Table Used to Calculate the Phi Coefficient

	Variable Y		
Variable X	0	1	Total
0	a	b	a + b
1	c	d	c + d
Total	a + c	b + d	N

The values for the phi coefficient for a 2 × 2 table can range from 0 to 1.00. When tables have dimensions that are larger than 2 × 2, phi may not lie between these two values. For that reason, the phi coefficient is restricted to the analysis of 2 × 2 tables and an extension of this statistic, the Cramér coefficient, is used for larger tables.

If the values of the two dichotomous variables have been coded "0" and "1," the phi coefficient and the Pearson product-moment correlation are identical except, possibly, for sign. For that reason, the strength of relationship between two such categorical variables is interpreted in a context similar to that for the Pearson r (Table 7.6) (Hinkle et al., 1994). That is, values of r_{pb} above .90 indicate an extremely strong relationship, .70 to .89 a strong relationship, .50 to .69 a moderate relationship, .30 to .49 a low relationship, and below .30 a weak relationship.

Critical Assumptions of the Phi Coefficient

Because the phi coefficient is determined by the chi-square statistic, the critical assumptions of the chi-square test for two independent samples (Chapter 7) apply here. The variables are dichotomous (2 × 2 tables), and observations are independent and consist of frequencies, not scores. The data in our hypothetical study partially meet all these assumptions, with the exception of random selection. Both the independent and dependent variables (group membership and children's fear) have been measured on a nominal scale with two levels.

Computer Commands

In SPSS for Windows, the phi coefficient is generated from the same computer commands used for the chi-square test of association (Chapter

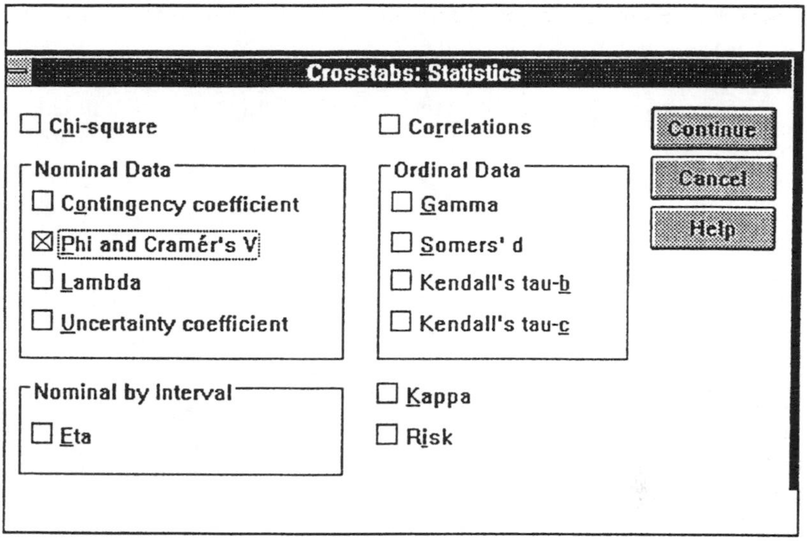

Figure 9.1. SPSS for Windows Computer Commands for the Phi Coefficient

8). To obtain the phi coefficient, open the *Crosstabs Statistics* dialogue box by clicking on *Statistics . . . Summarize . . . Crosstabs . . . Statistics*. Several statistical options, including the phi coefficient, are presented for the analysis of the association between nominal variables (Figure 9.1).

Computer-Generated Output

The syntax commands, results of the Fisher exact test, and requested phi coefficient for our data already have been presented in Table 7.5, but, for ease of discussion, they are duplicated in Table 9.3. The syntax commands indicate that the phi coefficient has been requested (Table 9.3, ①). The resulting analyses indicate that the obtained value of phi was .40 (②). Because the generated p value for the χ^2 statistic ($p = .028$, ③) is less than our prestated alpha (.05), this phi coefficient is significant. It could also be obtained as follows:

$$\varphi = \sqrt{\chi^2/N} = \sqrt{4.82/30} = .40$$

According to the Hinkle et al. (1994) criteria for assessing the strength of association (presented in Table 7.6), a phi coefficient of .40 is, at best, weak. Our conclusion would be, therefore, that there is a significant but weak association between group membership and postintervention fear.

Presentation of Results

The results of the statistical analysis using the phi coefficient could be presented along with results of the chi-square test of association (Table 7.7). They also could be presented in the text as follows:

> The results of the chi-square analysis indicate a significant but weak association between group membership (intervention, control) and postintervention fear ($\varphi = .40, p = .03$).

Advantages and Limitations of the Phi Coefficient

The phi coefficient offers a useful test for assessing the strength of relationship between two dichotomous variables when the chi-square test of association has been found to be significant. Because the phi coefficient takes into account sample size, it also allows the researcher to compare strengths of association across studies. A disadvantage to use of this statistic is that it can take on values greater than 1 when the table is larger than 2×2. For that reason, a related statistic, the Cramér V coefficient, is recommended for use with larger tables.

**Parametric and Nonparametric Alternatives
to the Phi Coefficient**

When the data from the independent and dependent dichotomous variables are coded "0" and "1," the phi coefficient is equivalent to the absolute value of the Pearson product-moment correlation coefficient. Nonparametric alternatives to the φ coefficient are the contingency coefficient (Chapter 7), the Cramér V coefficient (discussed below), and the Tetrachoric correlation (Hinkle et al., 1994). The Tetrachoric correlation is used when both dichotomous variables are assumed to have underlying continuity with a normal distribution (e.g., two items on a particular test).

Table 9.3 Computer-Generated Output From SPSS for Windows for the
Phi Coefficient

```
CROSSTABS
/TABLES=group  BY postfear
/FORMAT= AVALUE NOINDEX BOX LABELS TABLES
/STATISTIC=CHISQ CC PHI  ①
/CELLS= COUNT EXPECTED ROW COLUMN SRESID .

GROUP  exper-control grps  by  POSTFEAR  post intervention fear

                        POSTFEAR      Page 1 of 1
              Count  |
              Exp Val |
              Row Pct |not fear fearful
              Col Pct |ful                      Row
              Std Res |      .00|    1.00| Total
GROUP        --------+--------+--------+
         .00 |    10  |     5  |    15
  experimental |    7.0  |    8.0  | 50.0%
              |  66.7%  |  33.3%  |
              |  71.4%  |  31.3%  |
              |   1.1   |  -1.1   |
              +--------+--------+
        1.00 |     4  |    11  |    15
   control   |    7.0  |    8.0  | 50.0%
              |  26.7%  |  73.3%  |
              |  28.6%  |  68.8%  |
              |  -1.1   |   1.1   |
              +--------+--------+
       Column     14       16       30
       Total    46.7%    53.3%   100.0%

       Chi-Square          Value        DF   Significance
-----------------------  ----------     ----  ----------
Pearson                    4.82143        1    .02811  ③
Continuity Correction      3.34821        1    .06728
Likelihood Ratio           4.96252        1    .02590
Mantel-Haenszel test for   4.66071        1    .03086
linear association

Minimum Expected Frequency -    7.000

Approximate
Statistic               Value     ASE1  Val/ASE0  Significance
--------------------    --------- ------- -------- ------------
Phi                      .40089  ②                 .02811  ③
Cramer's V               .40089                    .02811
Contingency Coefficient  .37210                    .02811
```

Examples From Published Research

Bond, C. A., & Monson, R. (1984). Sustained improvement in drug documentation, compliance, and disease control: A four-year analysis of an ambulatory care model. *Archives of Internal Medicine, 144,* 1159-1162.

Eaves, R. C., & Milner, B. (1993). The criterion-related validity of the Childhood Autism Rating Scale and the Autism Behavior Checklist. *Journal of Abnormal Child Psychology, 21,* 481-491.

Krettek, J. E., Arkin, S. I., Chaisilwattana, P., & Monif, G. R. (1993). *Chlamydia trachomatis* in patients who used oral contraceptives and had intermenstrual spotting. *Obstetrics and Gynecology, 81*, 728-731.

Stolovitzky, J. P., & Todd, N. W. (1990). Head shape and abnormal appearance of tympanic membranes. *Otolaryngology, Head, & Neck Surgery, 102*, 322-325.

Cramér's *V* Coefficient

When the contingency table is greater than 2×2, an alternative measure of strength of association is Cramér's *V* coefficient (Cramér, 1946). In some texts (Daniel, 1990; Siegel & Castellan, 1988), the Cramér statistic is referred to as *Cramér's C* coefficient. Other texts (Hays, 1994; Hinkle et al., 1994) and SPSS for Windows label this statistic *Cramér's V*. For consistency with the computer printouts being reviewed, the Cramér statistic will be referred to in this text as *Cramér's V*. This is a modified version of the phi coefficient that adjusts for the number of levels of the categorical variable, thus allowing the coefficient to retain its range of 0 and 1. When the contingency table is 2×2, this statistic has the same value as the phi coefficient.

An Appropriate Research Question
for Cramér's *V* Coefficient

Cramér's *V* coefficient is useful when the researcher is interested in assessing the strength of association between two categorical variables once a χ^2 statistic has been determined to be significant. For example, Metheny and colleagues (1989) used Cramér's *V* after obtaining a significant chi-square to evaluate the effectiveness of pH measurements in predicting feeding tube placement in a sample of 181 adults. Herth (1990) used the same statistic to evaluate the contribution of various psychosocial factors to grief resolution in elderly widowers.

In Chapter 8, we used the chi-square test for *k* independent samples to examine the association between a family's social status (low, medium, high) and a mother's assessment of her child's posthospital adjustment (poor, moderate, excellent). In that example, a significant association was found between a family's social status and the child's adjustment. Given this significant result, we might be interested in assessing the strength of that significant relationship. Cramér's *V* coefficient can be of help with this analysis. A research question that would be suitable for use with Cramér's *V* is as follows:

Table 9.4 Example of Null and Alternative Hypotheses Suitable for Testing
With the Cramér Coefficient

Null Hypothesis
H_0: There is no association between the variables of family social status (low, medium,
or high) and mothers' perceptions of their children's posthospital adjustment (poor,
moderate, or excellent).

Alternative Hypothesis
H_a: There is an association between the variables of family social status (low, medium,
or high) and mothers' perceptions of their children's posthospital adjustment (poor,
moderate, or excellent).

What is the strength of the relationship between a family's social status (low,
medium, high) and a mother's assessment of her child's posthospital adjustment (poor, moderate, excellent)?

Null and Alternative Hypotheses

Table 9.4 presents null and alternative hypotheses that would be suitable
for use with the Cramér coefficient. Note that the hypotheses are similar to
those of the chi-square test for k independent samples (Table 8.1) because
Cramér's V is based on the χ^2 statistic.

Overview of the Procedure

The frequency data are first arranged in an $r \times c$ contingency table, and
a χ^2 statistic is computed (Chapter 8). From this obtained χ^2 value,
Cramér's V coefficient is obtained:

$$\text{Cramér's } V = \sqrt{\frac{\chi^2}{N(L-1)}}$$

where

L = the smaller of the number of rows or columns in the
contingency table

N = the total sample size.

Because Cramér's V is based on the χ^2 statistic, its significance level is the same as that for the χ^2; that is, the null hypothesis is rejected if the calculated χ^2 is greater than the critical χ^2 with $df = (r - 1)(c - 1)$ or, alternatively, if the generated p value for the χ^2 is less than the predetermined level of α.

Like the phi coefficient for 2×2 tables, the value of Cramér's V can range between 0 and 1, with higher values indicating greater strength of association. A limitation to this statistic is that although a value of 0 indicates no association between the two variables, a value of 1 does not always imply a perfect relationship (Daniel, 1990; Siegel & Castellan, 1988). This perfect relationship occurs only when the contingency table being analyzed is square (i.e., there are as many rows as columns in the table). If the contingency table has more rows than columns (or vice versa), a value of 1 for the Cramér coefficient would indicate that there is a perfect relationship in one direction (e.g., from the row to column variable) but not necessarily in the other (Siegel & Castellan, 1988). Siegel and Castellan also indicate that, except for 2×2 tables, values of Cramér's V are not directly comparable to the Pearson product-moment correlation; therefore, the guidelines offered for interpreting the strength of the phi coefficient (Table 7.6) do not necessarily apply to Cramér's V. Larger values of this coefficient, however, do indicate a greater degree of relationship between two categorical variables.

Critical Assumptions of Cramér's V Coefficient

Because Cramér's V coefficient is calculated from the χ^2 statistic, the requirements for this coefficient are similar to those for the χ^2 statistic for $r \times c$ independent groups outlined in Chapter 8; that is, assumptions are made that the variables being examined are categorical; the pairs of randomly selected observations are independent; the data being analyzed are frequency data, not scores; and the cells in the $r \times c$ contingency table are mutually exclusive and exhaustive. The selected data from our hypothetical study meet all these assumptions except for random selection.

Computer-Generated Output

Table 9.5 presents the syntax commands and computer-generated output that were obtained in SPSS for Windows. Cramér's V statistic was obtained by opening the *Crosstabs* window using the following commands: *Statistics . . . Summarize . . . Crosstabs* and selecting Cramér's V statistic from the *Statistics* dialogue box presented in Figure 9.1. Note that the syntax command *phi* will produce both the phi and Cramér coefficients (Table 9.5, ①).

In the printout, we are presented with the contingency table and chi-square values, the interpretation of which is presented in detail in Chapter 8. As requested, we are also presented with Cramér's *V* coefficient (②). This statistic was obtained as follows:

$$\text{Cramér's } V = \sqrt{\frac{\chi^2}{N(L-1)}} = \sqrt{\frac{9.54}{60(3-1)}} = .28$$

As indicated, this coefficient has the same significance level (.049) as the χ^2 statistic from which it was derived (③). Because .049 is less than our prestated α level (.05), we will reject the null hypothesis of no association and conclude that there is a significant association between social status and posthospital adjustment. Given that the Cramér statistic can range between 0 and 1, the strength of this relationship, .28, is weak.

Presentation of Results

The Cramér statistic can be presented in a table similar to Table 8.5, along with the frequencies and χ^2 statistic. It could also be presented in the text as follows:

The results of the chi-square analyses indicate a statistically significant but weak association between family social status and posthospital adjustment ($\chi^2 = 9.54$, $p = .049$, Cramér's $V = .28$).

Chapter 8 also suggests potentially useful approaches to determining which cells in the contingency table appear to have the most influence on these results.

Advantages and Limitations of Cramér's *V* Coefficient

A major advantage to Cramér's *V* statistic is that it does not have many assumptions associated with it. A disadvantage of this statistic is that, unlike the phi coefficient, the Cramér coefficient does not bear any direct relationship to the Pearson correlation coefficient if the contingency table is larger than 2 × 2. Moreover, if the table is not square, the value of 1 for

Table 9.5 Computer-Generated Output Obtained From SPSS for Windows for the Chi-Square Test for *k* Independent Samples

```
CROSSTABS
   /TABLES=socstat  BY adjust
   /FORMAT= AVALUE NOINDEX BOX LABELS TABLES
   /STATISTIC=CHISQ PHI   ①
   /CELLS= COUNT EXPECTED ROW COLUMN SRESID .

SOCSTAT  social status by  ADJUST  posthospital adjustment

                    ADJUST                    Page 1 of 1
             Count  I
             Exp Val I
             Row Pct IPoor    Moderate Excellent
             Col Pct I
             Std Res I   1.00I   2.00I   3.00I  Row
SOCSTAT      --------+--------+--------+--------+  Total
             1.00  I    6  I    7  I    3  I      16
     low           I   4.3  I   7.2  I   4.5  I   26.7%
                   I  37.5% I  43.8% I  18.8% I
                   I  37.5% I  25.9% I  17.6% I
                   I    .8  I   -.1  I   -.7  I
                   +--------+--------+--------+
             2.00  I    6  I   15  I    4  I      25
     middle        I   6.7  I  11.3  I   7.1  I   41.7%
                   I  24.0% I  60.0% I  16.0% I
                   I  37.5% I  55.6% I  23.5% I
                   I   -.3  I   1.1  I  -1.2  I
                   +--------+--------+--------+
             3.00  I    4  I    5  I   10  I      19
     high          I   5.1  I   8.6  I   5.4  I   31.7%
                   I  21.1% I  26.3% I  52.6% I
                   I  25.0% I  18.5% I  58.8% I
                   I   -.5  I  -1.2  I   2.0  I
                   +--------+--------+--------+
             Column    16      27      17        60
             Total   26.7%   45.0%   28.3%    100.0%


     Chi-Square               Value          DF    Significance
----------------------      ----------       ----  ------------
Pearson                      9.54490          4     .04883    ③
Likelihood Ratio             9.14369          4     .05761
Mantel-Haenszel test for     4.11150          1     .04259
   linear association

Cells with Expected Frequency < 5 -     2 OF    9 ( 22.2%)

Approximate
    Statistic               Value     ASE1   Val/ASE0 Significance
----------------------      --------- ------- -------- ------------
Phi                          .39885                      .04883
Cramer's V                   .28203  ②                   .04883   ③
```

Cramér's *V* does not imply a perfect relationship between the two categorical variables. Cramér's *V* nevertheless does allow the researcher to compare tables of different sizes and tables based on different sample sizes but with the same number of dimensions.

Parametric and Nonparametric Alternatives to Cramér's V Coefficient

If the contingency table being examined is 2×2 and the categories are coded "0" and "1," Cramér's V takes on the same value as the phi coefficient and shares the phi coefficient's relationship to the Pearson product-moment correlation, $V = \phi = |r|$. If the contingency table is larger than 2×2 and the categories are ordered and equally spaced, a nonparametric correlation such as Spearman's rho (discussed later in this chapter) might be considered. If there is no order to the categories, there is no parametric equivalent to Cramér's V and there are few nonparametric alternatives to this test.

Examples From Published Research

Herth, K. (1990). Relationship of hope, coping styles, concurrent losses, and setting to grief resolution in the elderly widow(er). *Research in Nursing & Health, 13*, 109-117.

Metheny, N., Williams, P., Wiersema, L., Wehrle, M. A., Eisenberg, P., & McSweeney, M. (1989). Effectiveness of pH measurements in predicting feeding tube placement. *Nursing Research, 38*, 280-285.

The Kappa Coefficient

Researchers in health care often make use of clinical observations of health care providers to evaluate client outcomes or to assign patients to diagnostic categories. It is important when undertaking such research to determine the extent to which the variability in the observed outcomes can be attributed to true variation in the characteristic that is being assessed or is, instead, the result of errors of measurement, including observer biases and disagreement. Researchers interested in partitioning out such variability would typically assess *interobserver agreement*, or the extent to which observers agree in their ratings or assignment to categories of the characteristic of interest.

Numerous approaches have been used to assess interobserver agreement. These approaches typically take the form of evaluating the percentage of agreement among observers. That is,

$$\text{percentage agreement} = \frac{\text{number of agreements}}{\text{number of agreements} + \text{number of disagreements}} \times 100$$

where

number of agreements	=	the number of cases rated the same by two raters
number of disagreements	=	the number of cases rated differently by two raters.

Although percentage agreement gives a rough estimate of interobserver agreement, a major problem with this statistic is that it does not consider the fact that some of the observer agreement can be expected to occur purely by chance (Bakeman & Gottman, 1986). An advantage to the kappa coefficient is that it takes into account chance agreement.

For example, two or more practitioners might be asked to assign each of n patients to one of k diagnostic or outcome categories. The extent of their agreement could then be evaluated to determine the extent to which the practitioners agree on the placement of the patients in the particular categories. If the diagnostic or treatment variables to which the patients being assigned are at the nominal level of measurement and if the researcher also wants to adjust for chance agreement, the kappa coefficient (Cohen, 1960) can used to assess the extent of interobserver or interrater agreement.

The kappa coefficient (κ) represents a family of indexes of agreement that vary slightly from one another in their form and function. Extensive discussions of the various forms that these kappas can take can be found in Cohen (1960, 1968), Fleiss (1971), and Bakeman and Gottman (1986). Additional discussion regarding the use of the kappa coefficient can be found in Blackman and Koval (1993), Brennan and Hays (1992), Brennan and Prediger (1981), and Green (1981). The discussion in this chapter will focus on a very basic form of kappa, the value of which can be obtained in SPSS for Windows.

An Appropriate Research Question for the Kappa Coefficient

The kappa coefficient has been used in a variety of studies. Beach and colleagues (1992) used the kappa coefficient to evaluate the degree of agreement between spouses and patients recovering from acute myocardial infarctions concerning the couples' reporting of frequency of sexual activity. Jacobson and Moore (1981) used a similar approach to examine the reliability of spouses as observers of behaviors that occur in their marital relationships.

In our hypothetical study, suppose we were interested in comparing our two groups of children (intervention and control) with regard to nurse evaluations of their postoperative recovery (poor, moderate, excellent) immediately following surgery. A convenient and economical approach would be to assign a single observer the task of evaluating a child's recovery. A serious drawback to this approach is that it would not be possible to determine whether the variability in recovery observed among the children in the two groups was a result of true differences among the children or instead resulted from observer bias. To assess the extent of this potential bias, we would want to include an additional observer of the children's recovery for at least a portion of the sample to determine the extent of interobserver agreement.

Assume, therefore, that two nurses were asked to assess 30 of the participating children's postoperative recovery immediately following surgery. A research question that could be answered using the kappa coefficient would be:

To what extent do the two nurses agree in their assessment of children's postoperative recovery immediately following surgery?

Null and Alternative Hypotheses

Table 9.6 presents examples of possible null and alternative hypotheses generated from our research question that would be suitable for use with the kappa coefficient. Because we are interested in the extent of *agreement*, not *disagreement*, between two raters, the research hypothesis is directional. We are testing the null hypothesis H_0: $\kappa \leq 0$ against the one-tailed alternative hypothesis H_a: $\kappa > 0$. The reason for this directional test is that, in our research hypothesis, we are addressing *beyond chance* agreement and are not contemplating the depressing possibility that there could be *less than chance* agreement between the raters. The null hypothesis of no interobserver agreement beyond chance will be rejected if the significance of kappa is greater than the critical value of a z statistic at one-tailed alpha $= .05$ (i.e., critical $z = 1.64$).

Table 9.6 Example of Null and Alternative Hypotheses Appropriate for Use With the Kappa Statistic

---

Null Hypothesis

H_0:   The two nurses do not agree on their evaluations of children's postoperative recovery (poor, moderate, or excellent).

Alternative Hypothesis

H_a: The two nurses do agree on their evaluations of children's postoperative recovery (poor, moderate, or excellent).

Overview of the Procedure

The basic form that kappa takes is as follows (Bakeman & Gottman, 1986):

$$\kappa = \frac{\text{proportion of observed agreement} - \text{proportion of chance agreement}}{1 - \text{proportion of chance agreement}}$$

$$= \frac{P_o - P_c}{1 - P_c}$$

Given a square $r \times c$ contingency table, the proportion of observed agreement (P_o) is determined by summing the number of agreements that appear on the diagonal of the contingency table and dividing by the total number of paired observations (i.e., N = the number of observer agreements + the number of disagreements):

$$P_o = \frac{\text{number of agreements}}{N}.$$

The proportion of chance agreement (P_c) for each cell on the diagonal is determined in a manner similar to that used for the chi-square statistic in Chapter 7. First, the row and column marginal totals for each cell on the diagonal are multiplied together; the result is divided by the total number of observations. This provides the number of observations that could have been expected to occur by chance in the particular cell on the diagonal. The proportion of chance agreement, P_c, is obtained by dividing this expected frequency by the total number of observations. Finally, these proportions are summed across all the cells on the diagonal to obtain the total proportion of chance agreement:

$$P_c = \sum_{i=1}^{k} \frac{(\text{row marginal})\,(\text{column marginal})}{N^2}.$$

Given these two proportions, all that is needed is to plug in the values into the kappa formula. We will do this for our hypothetical study when we examine the computer output.

Values of kappa can theoretically range from -1 to $+1$. A negative value of kappa implies that the proportion of agreement resulting from chance is greater than the proportion of observed agreement, not a desirable situation. For that reason, alternative hypotheses are directional, with higher positive values of kappa indicating stronger interobserver agreement.

Assessing the Kappa Coefficient

There are two approaches to assessing a kappa coefficient: testing its level of significance (Bakeman & Gottman, 1986) and evaluating its magnitude, using criteria suggested by Fleiss (1971). Siegel and Castellan (1988) indicate that, for large N, kappa is approximately normally distributed. To determine whether the obtained kappa is significantly greater than 0, the obtained kappa is divided by its standard error (i.e., the square root of its estimated variance) to produce a z statistic that is used for hypothesis testing:

$$\text{significance of } \kappa = z = \frac{\kappa}{\sqrt{\text{Var}\,(\kappa)}}.$$

The null hypothesis that $\kappa \leq 0$ will be rejected if the value of this generated z statistic is greater than the critical value of z at the prestated one-tailed level of alpha (e.g., $z = 1.64$ at $\alpha = .05$).

The estimated variance of kappa is somewhat involved to calculate directly from a contingency table. Fortunately, SPSS for Windows presents the value of this z statistic (Table 9.7, ④). Readers interested in details of the procedure used to calculate the estimated variance of kappa are referred to Bakeman and Gottman (1986) and Siegel and Castellan (1988). The interpretation of the significance of kappa will be discussed in greater detail when we examine the computer printout.

Although a significant z statistic indicates that the observers agree significantly more than would be expected by chance, this statistic does not indicate the strength of agreement. To assess this situation, the researcher needs to examine the size of kappa itself. Unfortunately, there are no strict guidelines with which to assess values of kappa. Bakeman and Gottman (1986) view kappas that are less than .70 with some concern. This may be a bit stringent, however. Fleiss (1971), for example, indicates that kappas of .40 to .60 are fair, .60 to .75 are good, and values greater than .75 are excellent.

Critical Assumptions of the Kappa Coefficient

The critical assumptions for the kappa coefficient are as follows:

1. *The nominally scaled data are paired observations of the same phenomena (e.g., Observer 1 vs. Observer 2).*
2. *Observations are assigned to categories that are mutually exclusive and may or may not have order to them.*
3. *The resulting agreement or "confusion matrix" (Bakeman & Gottman, 1986) is symmetric (same number of rows and columns), such as 2 × 2 or 3 × 3.*

The data from our hypothetical study meet all these assumptions except for random selection. The two nurses both evaluated the postoperative recovery of the same group of 30 children. The categories of postoperative recovery (poor, moderate, and excellent) were mutually exclusive, ordered categories. The nurses' ratings resulted in a 3 × 3 agreement matrix that is evaluated in Table 9.7.

Computer-Generated Output

Table 9.7 presents the syntax commands (①) and computer-generated output from SPSS for Windows for the kappa coefficient. This printout can be obtained in SPSS for Windows using the same commands as for Cramér's V coefficient. That is, the *Crosstabs* window is first opened by highlighting the commands *Statistics . . . Summarize . . . Crosstabs* in the menu. Next, the kappa coefficient is selected from the *Statistics* dialogue box (Figure 9.1).

Table 9.7 Computer-Generated Printout From SPSS for Windows for the
Kappa Statistic

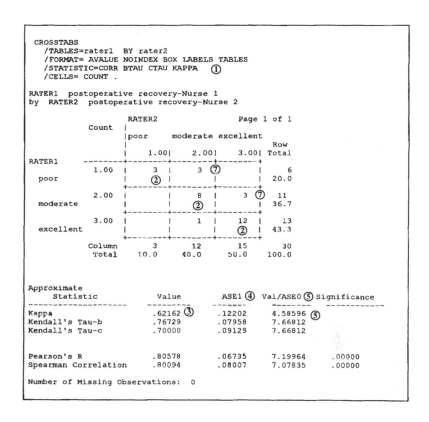

The syntax commands indicate that, as requested, the kappa coefficient
will be generated (①). The generated contingency table indicates the
frequency of agreement between the nurses with regard to the postoperative
recovery of the 30 children. The table also presents the row and column
totals and their percentages. This agreement or "confusion matrix" is a 3 ×
3 square contingency table in which the diagonal values (3, 8, and 12) (②)

represent the number of exact agreements between Nurse Observer 1 and Nurse Observer 2. Using the formulas outlined previously, the proportion of observed agreement, P_o, would be calculated as follows:

$$P_o = \frac{3 + 8 + 12}{30} = .7667$$

Next, the total proportion of chance agreement, P_c, is obtained for each of the three diagonal cells by multiplying the row and column marginals for each of these cells together and dividing by N^2:

$$P_c = \frac{(3)(6)}{30^2} + \frac{(12)(11)}{30^2} + \frac{(15)(13)}{30^2} = .3833 .$$

The probability that the two observers would agree purely by chance is .3833. Using the proportions of observed and chance agreement, the kappa coefficient (③) is easily calculated:

$$\kappa = \frac{P_o - P_c}{1 - P_c} = \frac{.7667 - .3833}{1 - .3833} = .6216 .$$

The interobserver agreement has now been corrected for chance agreement and is considerably lower (.6216) than the exact agreement (.7667). Bakeman and Gottman (1986) point out that these two values can be quite disparate if there are few coding categories and low marginal frequencies.

The next steps are to determine whether the obtained kappa is significantly greater than zero and to evaluate the size of the kappa coefficient. As indicated earlier, the z statistic that is used to evaluate whether the obtained kappa is significantly different from zero is:

$$\text{significance of } \kappa = z = \frac{\kappa}{\sqrt{\text{Var} (\kappa)}}$$

Unfortunately, we are not given the estimated value of the variance of kappa in the computer printout in SPSS for Windows. Although it would have been logical to provide it, the ASE1 that is shown in the printout (④) *is not* the standard error of kappa that is used in testing the null hypothesis, H_0: $\kappa \leq 0$ (Norusis, 1995a). Fortunately, the value that *is* presented under the heading Val/ASE0 (⑤) *is* the calculated value of our z statistic necessary to assess the significance of kappa (4.58). For the interested reader, Bakeman and Gottman (1986) present a step-by-step guide through the somewhat tedious calculation of this standard error of the variance. Had we undertaken these steps for these data, the resulting value for Var(κ) would have been .0184, resulting in a z statistic similar to that presented in the computer printout (⑤):

$$\text{significance of } \kappa = z \;=\; \frac{\kappa}{\sqrt{\text{Var}(\kappa)}} = \frac{.6216}{\sqrt{.0184}} = 4.58 .$$

Given that the resulting value for the significance of kappa, 4.58, is greater than the critical value of a z statistic at a one-tailed alpha of .05 (1.64), we can reject the null hypothesis that kappa ≤ 0. Our conclusion is, therefore, that the nurse observers significantly agreed with one another beyond chance levels with regard to their evaluations of the children's postoperative recovery. The size of this kappa, .6216, fits Fleiss's (1971) characterization of interobserver agreement as "fair to good."

Assessing Asymmetric Tables

Note that the table presented in the printout in Table 9.7 is square: It has equal numbers of rows and columns. This allows us to use the standard *integer* mode for the *Crosstabs* command in SPSS for Windows. Suppose, however, that one of the marginal frequencies had been equal to zero (e.g., Rater 2 did not consider any of the children to have poor postoperative recovery, whereas Rater 1 did). If such a situation were to occur and the *Crosstabs* command had been selected from the menus, the computer would be unable to calculate the kappa coefficient because, in the *integer* mode (a default for the *Crosstabs* dialogue box), the table is asymmetric.

Should this asymmetric matrix occur—and it commonly does in the real world—it will be necessary to switch to the syntax mode in SPSS for Windows in order to operate the *general* mode for *Crosstabs*. This means that all the *Crosstabs* syntax commands (Table 9.7, ①) need to be typed into the syntax box. In addition, prior to the "/Tables" command, the variables and their potential ranges (e.g., "/Variables = rater1 (1,3) rater2 (1,3)") need to be specified followed by the remaining commands.

Table 9.8 gives an example of the syntax commands and resulting printout when one of the row or column marginals is equal to zero. Note

Table 9.8 Example of SPSS for Windows Syntax Commands and Output
When a Row or Column Marginal Is Equal to Zero

```
CROSSTABS  VARIABLES = rater1 (1,3) rater2 (1,3) ①
   /TABLES=rater1  BY rater2
   /FORMAT= AVALUE NOINDEX BOX LABELS TABLES
   /STATISTIC=KAPPA
   /CELLS= COUNT.

RATER1  postoperative recovery-Nurse 1
by  RATER2  postoperative recovery-nurse 2

                        RATER2
                Count  |
                       |moderate excellent
                       |                       Row
                       |   2  |   3  | Total
RATER1          -------+------+------+
                1  |      6  |   0  |     6
         poor      |         |      |    20.0
                   +---------+------+
                2  |      8  |   3  |    11
       moderate    |         |      |    36.7
                   +---------+------+
                3  |      1  |  12  |    13
      excellent    |         |      |    43.3
                   +---------+------+
                Column    15       15        30
                Total   50.0     50.0     100.0

Approximate
  Statistic                 Value        ASE1      Val/ASE0    Significance
------------------------  ---------    ---------    --------    ------------
Kappa                       .44444       .12094     3.27510

Number of Missing Observations:   0
```

that commands have changed slightly (①) to accommodate the apparent
asymmetry in the table and to indicate to SPSS for Windows that we are in
the *general* mode for *Crosstabs*.

Presentation of Results

Table 9.9 is an example of a suitable presentation of the results of the
analysis of interobserver agreement for the computer output presented in
Table 9.7. It is also possible to present the results in the text as follows:

An evaluation of agreement between two nurse observers with regard to their
assessments of 30 children's postoperative recovery was undertaken. The exact
percentage agreement between the two observers was .77. A kappa coefficient

Table 9.9 Presentation of Results of Interobserver Agreement on Children's Postoperative Recovery Using the Kappa Coefficient

	Nurse Observer 2			
Nurse Observer 1	Poor	Moderate	Excellent	Total
Poor	3	3	0	6
Moderate	0	8	3	11
Excellent	0	1	12	13
Total	3	12	15	30

was used to correct for chance agreement among the observers. This resulted in a significant kappa of .62 ($z = 4.59$, $p < .0001$). The size of this kappa indicates that there was fair to good interobserver agreement (Fleiss, 1971).

Advantages and Limitations of the Kappa Coefficient

The kappa coefficient is an extremely useful nonparametric statistic that can be used to evaluate interobserver agreement. One of its advantages is that the agreement matrix that is generated for this statistic graphically provides a clear display of agreements and disagreements among observers. This is especially useful for training purposes. For example, the matrix can illustrate which observers may be more sensitive to the nuances in the coding or diagnostic scheme. The agreement matrix can also point out where observer disagreements most often occur. As indicated in the computer printout in Table 9.7, six of the seven disagreements between Rater 1 and Rater 2 were situations in which Rater 1 evaluated the children lower with regard to postoperative recovery than did Rater 2 (⑦). These two cells were also where the greatest discrepancies occurred between the two observers. Future training, therefore, might be directed toward achieving greater consensus among the raters concerning the meaning of these categories.

A serious disadvantage to the kappa coefficient is that it is very sensitive to low values in the marginals and small contingency tables because chance values are higher for smaller contingency tables (Bakeman & Gottman, 1986). It also is limited to square contingency tables. As indicated, in SPSS for Windows, if a row or column total is 0, it is necessary to switch to syntax commands and the general mode for the *Crosstabs* command to obtain a kappa coefficient.

Parametric and Nonparametric Alternatives to the Kappa Coefficient

A number of refinements can be made to the kappa coefficient. For example, a weighted kappa (Cohen, 1968) can be used when the researcher considers some mistakes to be more important than others. This type of kappa weights disagreements differently, depending on how serious the researcher considers a particular error to be. Modified versions of the simple kappa coefficient have also been derived to assess agreement among more than two observers (Fleiss, 1971; Uebersax, 1982) and for use with interval-level data (Berry & Mielke, 1987). In addition, Feingold (1992) discusses the equivalence of kappa and Pearson's chi-square statistics in the 2×2 table.

A very useful parametric alternative to the use of percentage agreement in general, and kappa in particular, to assess interobserver agreement is to utilize a generalizability theory approach (Cronbach, Gleser, Nanda, & Rajaratman, 1972; Shavelson & Webb, 1991) to decompose the various errors of measurement into their appropriate components. Shavelson and Webb (1991), in particular, offer a very clearly written overview for nonmathematicians of the theory and methods of generalizability.

Examples From Published Research

Beach, E. K., Maloney, B. H., Plocica, A. R., Sherry, S. E., Weaver, M., Luthringer, L., & Utz, S. (1992). The spouse: A factor in recovery after acute myocardial infarction. *Heart & Lung: The Journal of Critical Care, 21,* 30-38.

Jacobson, N. S., & Moore, D. (1981). Spouses as observers of the events in their relationship. *Journal of Consulting and Clinical Psychology, 49,* 269-277.

The Point Biserial Correlation

The phi, Cramér, and kappa coefficients are suitable for assessing the strength of association in categorical data. There may be occasions, however, when researchers are interested in assessing the strength of relationship between one variable that is at the interval or ratio level of measurement and a second variable that is either dichotomous or at the ordinal level of measurement (e.g., gender vs. depression scores, or one scale item vs. the total scale score). The *point biserial correlation* is used to examine strength of association in the first instance (dichotomous vs. interval/ratio data), and the *biserial correlation* is used in the second instance (ordinal vs. interval data). In this section, we will examine the point biserial correlation in greater detail; the biserial correlation will be briefly addressed when examining the Spearman rho correlation coefficient.

An Appropriate Research Question for the Point Biserial Correlation

The point biserial correlation has been used in a number of health care research projects. Damrosch and Perry (1989) used the point biserial correlation in their analysis of patterns of adjustment, frequency of chronic sorrow, and coping behaviors of parents of children with Down syndrome. Knapp-Spooner and Yarcheski (1992) used the same point biserial correlation to examine the use of sleep medication and sleep disturbance in 24 adult patients undergoing coronary artery bypass graft surgery. Another typical use of the point biserial correlation is to compare an individual's performance on a single item (e.g., yes, no) with his or her total score on a particular test.

In our hypothetical study, a number of possible research questions could be answered using the point biserial correlation. For example, suppose we were interested in examining the strength of association between a child's gender and his or her posthospital adjustment. A research question that could be answered using this statistic would be as follows:

What is the strength of association between a child's gender (male, female) and his or her posthospital adjustment scores?

Null and Alternative Hypotheses

Table 9.10 presents null and an alternative hypotheses that were generated from our research question and that could be answered using a point biserial correlation. Note that the variable of gender is nominal (0 = male, 1 = female) and that the posthospital adjustment variable is at the interval level of measurement, with scores that range from 62 to 93. Higher scores indicate more satisfactory posthospital adjustment.

Table 9.10 Example of Null and Alternative Hypotheses Suitable for a
Point Biserial Correlation Coefficient

Null Hypothesis
H_0:   There is no association between between the variables of a child's gender (male, or
female) and his or her posthospital adjustment scores.

Alternative Hypothesis
H_a: There is an association between the variables of a child's gender (male or female) and
his or her posthospital adjustment scores.

Overview of the Procedure

The point biserial correlation is a special case of a parametric statistic,
the Pearson product-moment correlation. Although there may be some
disagreement in the statistics literature as to whether the point biserial
correlation is a parametric or nonparametric statistic, it is considered by
some writers (e.g., Daniel, 1990) to be nonparametric because one of the
variables being assessed is dichotomous.

Calculation of the point biserial correlation, r_{pb}, is rather straightforward
and is merely a simplified version of the Pearson r:

$$r_{pb} = \sqrt{\frac{n_1\, n_o}{N}}\left(\frac{(\bar{x}_1 - \bar{x}_o)}{\sqrt{\sum (x - \bar{x})^2}}\right)$$

where

N = total sample size

n_0 = the number of cases when the value of the
dichotomous variable = 0

n_1 = the number of cases when the value of the
dichotomous variable = 1

$\bar{x}_0$ = mean of the continuous variable when the
dichotomous variable = 0

$\overline{x}_1$ = mean of the continuous variable when the
dichotomous variable = 1

x = individual observations for entire group

$\overline{x}$ = the mean for the entire group.

The value of the point biserial correlation coefficient can range between -1 and $+1$, with higher absolute values indicating a greater strength of relationship. The point biserial correlation is tested in a manner similar to that for the Pearson r. The null hypothesis of no relationship between the dichotomous and continuous variables will be rejected if the obtained value of the correlation coefficient exceeds a critical value with $df = N - 2$ at a prestated alpha level (e.g., .05), or, alternatively, if the resulting p value is less than the prestated alpha.

Because of the close relationship of r_{pb} to the Pearson r, the strength of association for r_{pb} can be evaluated using the criteria for r^2 (Table 7.6) (Hinkle et al., 1994). That is, values of r_{pb}^2 above .81 indicate an extremely strong relationship, .49 to .80 a strong relationship, .25 to .48 a moderate relationship,.09 to .24 a low relationship, and .00 to .08 a weak relationship.

Critical Assumptions of the
Point Biserial Correlation

The point biserial correlation has a number of assumptions that are associated with both nonparametric and parametric statistics.

1. *The randomly selected observations are paired, with one variable being dichotomous and the other at the interval/ratio level.*
2. *The dichotomous (Y) variable has mutually exclusive groups whose values have been coded 0 and 1.*
3. *The continuous (X) variable is distributed normally overall and within each level of the dichotomous variable.*
4. *The continuous (X) variable has equal variances across each level of the dichotomous variables.*

Note that all four of these assumptions are similar to those of the parametric independent (or Student's) t test. The difference in the two tests is that whereas the independent t test is a test of differences in the means of two groups, the point biserial correlation is a measure of association between two variables. Moreover, whereas the independent variable for the *t* test *must* be dichotomous, there is no such restriction for the point biserial correlation. Either the independent or dependent variable may be dichotomous.

In our hypothetical study, the independent variable, gender, is dichotomous and the dependent variable, posthospital adjustment, is at the interval level. The *Statistics . . . Summarize . . . Explore* dialogue box in SPSS for Windows can be used to generate the plots and K-S statistics to determine whether we meet the assumption of normality of distribution for the continuous variable across both levels of the dichotomous variable (see Chapter 3 for details).

Homogeneity of variance could also be assessed in SPSS for Windows by opening the *Statistics . . . Compare Means . . . Independent-Samples T-Test* dialogue box in the menu. The *grouping variable* is the dichotomous variable (e.g., gender), and the *test variable* is the continuous variable (e.g., posthospital adjustment). The resulting independent *t* test also provides information about the *Levene's test for equality of variances*, which is a test of homogeneity of variance. The null hypothesis of equal variances will be rejected if the generated *p* value of the Levene's test is less than our alpha (e.g., $\alpha = .05$). For this test, we do not want to reject H_0 because that would imply that the variances are not equal.

Computer-Generated Output

Table 9.11 presents the syntax commands and computer-generated output for the point biserial correlation obtained in SPSS for Windows. SPSS for Windows does not produce a point biserial correlation directly. Because this statistic is a special case of the Pearson correlation coefficient, it is possible to obtain the point biserial correlation by opening the *Statistics . . . Correlate . . . Bivariate* dialogue box, identifying both the nominal- and interval-level variables of interest (i.e., gender and posthospital adjustment), and selecting the *Pearson* correlation coefficient.

Table 9.11 indicates that the point biserial correlation coefficient generated from SPSS for Windows is –.48 (①). This value could also have been obtained directly from the formula for the point biserial correlation if we had been given the following information concerning sample sizes, group means, and the total variance for the continuous variable.

Table 9.11 Computer-Generated Printout From SPSS for Windows for the Point Biserial Correlation

```
CORRELATIONS
  /VARIABLES=gender posthosp
  /PRINT=TWOTAIL SIG
  /MISSING=PAIRWISE .


                    - - Correlation Coefficients  - -

              GENDER      POSTHOSP

GENDER        1.0000       -.4855  ①
             (   30)      (   30)
             P= .         P= .007  ②

POSTHOSP      -.4855       1.0000
             (   30)      (   30)
             P= .007      P= .

(Coefficient / (Cases) / 2-tailed Significance)

" . " is printed if a coefficient cannot be computed
```

N $= 30$

n_0 $= 15$ (the number of cases in the intervention group for which gender $= 0$)

n_1 $= 15$ (the number of cases in the control group for which gender $= 1$)

s^2 $= \Sigma(X - \overline{X})^2/(N - 1) = 99.70$

$\Sigma(X - \overline{X})^2 = (N - 1)\, s^2 = (30 - 1)(99.70) = 2,891.30$

$\overline{X}_0$ $= 84.33$ (the mean value of posthospital adjustment for the males)

$\overline{X}_1$ $= 74.80$ (the mean value of posthospital adjustment for the females).

It is possible to generate the value for the point biserial correlation:

$$r_{pb} = \sqrt{\frac{n_1 \, n_o}{n}} \left(\frac{(\overline{X}_1 - \overline{X}_o)}{\sqrt{\sum (romanX - \overline{X})^2}} \right)$$

$$= \sqrt{\frac{(15)(15)}{30}} \left(\frac{(74.80 - 84.33)}{\sqrt{(2891.30)^2}} \right) = (2.739)(-.1772) = -.48.$$

Is the value of the point biserial correlation, −.48, sufficiently large to reject the null hypothesis of no relationship between the variables of gender and posthospital adjustment? To determine this, we need to examine the p value that is presented in the printout ($p = .007$, ②). We will reject the null hypothesis if the generated p value is less than our predetermined two-tailed alpha (e.g., $\alpha = .05$). Because .007 is less than .05, we will reject the null hypothesis of no association and conclude that there is a significant relationship between gender and posthospital adjustment. If our research hypothesis had been directional, we would either have chosen a one-tailed test of significance or have divided the resulting p value in half and compared it to our one-tailed α level. The inverse correlation that we have obtained suggests that the group assigned the value of 0 (males) had higher posthospital adjustment scores than the group assigned the value of 1 (females). According to our guidelines regarding strength of association, the relationship between gender and posthospital adjustment is moderate because $r_{pb}^2 = (-.48)^2 = .23$.

Presentation of Results

Although the point biserial correlation could be presented in a table (see Table 9.15), it is probably best suited to presentation in the text if there is only one biserial correlation to report. That is,

A point biserial correlation was undertaken to examine the strength of association between the categorical variable, gender, and the continuous variable, posthospital adjustment. The results of this analysis ($r_{pb} = -.48$, $p = .007$) indicate that, although the boys had significantly higher posthospital adjustment scores than the girls, the relationship between gender and posthospital adjustment was moderate. That is, 23% of the variance in the children's posthospital adjustment scores can be explained by gender.

Advantages and Limitations of the Point Biserial Correlation

The point biserial correlation is used to examine the strength of relationship between a nominal-level and an interval-level variable. It would be an especially useful statistic, therefore, when the researcher has found a significant difference between two groups (e.g., intervention vs. control) and wants to know how strong the relationship is between this nominal-level independent variable and a continuous dependent variable. Neither the *t* test nor any of its nonparametric alternatives (e.g., the Mann-Whitney test) can provide this information.

A disadvantage to this statistic is that it must meet the parametric assumption of a normally distributed continuous variable. The categorical variable of interest must also be dichotomous, with assigned values of 0 and 1.

Parametric and Nonparametric Alternatives to the Point Biserial Correlation

As indicated, the point biserial correlation is a special case of the Pearson product-moment correlation. An alternative nonparametric measure for ascertaining strength of association between a dichotomous and a continuous variable was suggested by Freeman (1965). This statistic has been found to have a distribution very similar to the Mann-Whitney *U* statistic (Buck & Finner, 1985; Daniel, 1990) but is, unfortunately, not available in SPSS for Windows.

Examples From Published Research

Damrosch, S. P., & Perry, L. A. (1989). Self-reported adjustment, chronic sorrow, and coping of parents of children with Down syndrome. *Nursing Research, 38*, 25-30.

Knapp-Spooner, C., & Yarcheski, A. (1992). Sleep patterns and stress in patients having coronary bypass. *Heart & Lung: The Journal of Critical Care, 21*, 342-349.

The Spearman
Rank-Order Correlation Coefficient

The Spearman rank-order correlation coefficient (also known as *Spearman's rho* or r_s) (Spearman, 1904) is one of the best-known and most frequently used nonparametric statistics in health care research (Pett & Sehy, 1996). Spearman's rho is used to examine the relationship between two ordinal-level variables. It is also a very suitable alternative to its parametric alternative, the Pearson product-moment correlation coefficient, when for various reasons the data do not meet that test's assumptions.

An Appropriate Research Question for the Spearman Rank-Order Correlation Coefficient

Numerous studies have used Spearman's rho to evaluate the degree of association between the rankings of two continuous variables. For example, Coyle and Sokop (1990) used a Spearman rank-order correlation coefficient to examine the extent of adoption of innovative practices among 113 hospital-based nurses. Cook, Smith, and Truman (1994) also used Spearman's rho to examine the relationship between patterns of recovery and length of stay in a rehabilitation unit for 53 brain-injured patients. Wells and Horm (1992) used this same statistic to examine the relationship between stage of disease at diagnosis and various racial and socioeconomic factors among female cancer patients.

In our hypothetical study, we might be interested in examining the relationship between the children's posttest anxiety scores and mothers' evaluations of their children's posthospital adjustment. The anxiety scores were measured on a 7-point Likert-type scale (1 = *not at all anxious*, 7 = *very anxious*) and thus were at the ordinal level of measurement. Although the posthospital adjustment scores were at the interval level of measurement, with a range of 62 to 93, they were not normally distributed, a requirement of the parametric Pearson product-moment correlation. A research question that could be answered using the Spearman rank-order correlation coefficient is as follows:

What is the relationship between children's posttest anxiety scores and mothers' perceptions of their children's posthospital adjustment?

Null and Alternative Hypotheses

An example of null and alternative hypotheses that are based on our research question and would be suitable for a Spearman rank-order correlation coefficient is presented in Table 9.12. Because our research question is directional, the alternative hypothesis is also directional.

Table 9.12 Example of Null and Alternative Hypotheses Suitable for a Spearman Rank-Order Correlation Coefficient

Null Hypothesis
H_0:   There is no association between children's posttest anxiety and mothers' perceptions of their children's posthospital adjustment.
Alternative Hypothesis
H_a: There is a negative association between children's posttest anxiety and mothers' perceptions of their children's posthospital adjustment.

Overview of the Procedure

Like the point biserial correlation coefficient, Spearman's rho is a special case of the Pearson r. For Spearman's rho, however, a Pearson correlation coefficient is obtained for the rankings of the observations, not their actual scores.

To compute r_s, the observations for each of the X and Y variables are ranked for each subject from lowest (rank = 1) to highest (rank = N). Tied observations are assigned the average rank that would have been assigned without ties. For example, if three observations on one variable were tied for third position and would have occupied positions 3, 4, and 5, they would all receive the rank of 4 (i.e., $[3 + 4 + 5]/3 = 4$). Next, for each subject, the difference between his or her rank on the X and Y variables, d_i, is obtained, squared, and summed across all subjects. If there are no ties, the Spearman rho correlation coefficient can be obtained using the following formula:

$$r_s = 1 - \frac{6 \sum_{i=1}^{N} d_i^2}{N^3 - N}$$

where

d_i = the difference in the ranks on the two paired variables, X and Y, for a particular subject

N = the number of pairs of observations.

This formula does not adjust for ties. Ties appear to have a minimal effect on the value of r_s provided there are few ties or the number of ties within a group of ties is small. The following formula for r_s adjusts for ties (Daniel, 1990; Siegel & Castellan, 1988):

$$r_s = \frac{(N^3 - N) - 6 \sum d_i^2 - (T_x + T_y)/2}{\sqrt{(N^3 - N)^2 - [(T_x + T_y)(N^3 - N)] + T_x T_y}}$$

The values for d_i and N are the same as for the formula presented previously. The correction factors for tied ranks, T_x and T_y, are obtained using the following formula (Siegel & Castellan, 1988):

$$T_x = \sum_{i-1}^{g} (t_i^3 - t_i)$$

where

g = the number of groupings of different tied ranks

t_i = the number of tied ranks in the ith grouping.

With these formulas, we could calculate Spearman's rho for the children's postintervention anxiety and posthospital adjustment scores by hand using the raw data presented in Table 9.13. Because there are so many ties for both the independent and dependent variables, it will be necessary to use the Spearman rho formula that corrects for ties; therefore, we need to compute not only d_i^2 (the difference in the rankings of the X and Y variables, Table 9.13, ①) but also T_x and T_y:

$$T_x = \sum_{i=1}^{g} (t_i^3 - t_i) = (4^3 - 4) + (7^3 - 7) + (11^3 - 11) + (6^3 - 6) + (2^3 - 2)$$

$$= 60 + 336 + 1320 + 210 + 6 = 1932$$

Table 9.13 Raw Data for Spearman Rho and Kendall's Tau-b Calculations

Adjust (Y)	Adjust (Y)	R_x	R_y	d	d^2 ①	# Natural Order Pairs-Concordant (C)	# Reverse Order Pairs-Discordant (D)	# Ties (T)
3 ②	76 ③	2.5	11	-8.5	72.25	16 ④	9 ⑤	1 ⑥
3	76	2.5	11	-8.5	72.25	16	9	1
3	85	2.5	19.5	-17.0	289.00	9	15	2
3	85	2.5	19.5	-17.0	289.00	9	15	2
4	75	8	9	-1.0	1.00	11	8	0
4	85	8	19.5	-11.5	132.25	5	14	0
4	85	8	19.5	-11.5	132.25	5	14	0
4	89	8	23.5	-15.5	240.25	4	15	0
4	89	8	23.5	-15.5	240.25	4	15	0
4	90	8	25.5	-17.5	306.25	4	15	0
4	90	8	25.5	-17.5	306.25	4	15	0
5	62	17	1	16.0	256.00	8	0	0
5	67	17	7	10.0	100.00	4	4	0
5	67	17	7	10.0	100.00	4	4	0
5	67	17	7	10.0	100.00	4	4	0
5	79	17	14	3.0	9.00	2	6	0
5	82	17	15	2.0	4.00	2	6	0
5	83	17	16.5	.5	.25	1	6	1
5	92	17	27.5	-10.5	110.25	0	8	0
5	92	17	27.5	-10.5	110.25	0	8	0
5	93	17	29.5	-12.5	156.25	0	8	0
5	93	17	29.5	-12.5	156.25	0	8	0
6	66	25.5	4.5	21.0	441.00	0	2	0
6	66	25.5	4.5	21.0	441.00	0	2	0
6	76 ⑥	25.5	11	14.5	210.25	0	2	0
6	78	25.5	13	12.5	156.25	0	2	0
6	83	25.5	16.5	9.0	81.00	0	2	0
6	86	25.5	22	3.5	12.25	0	2	0
7	65	29.5	2.5	27.0	729.00	0	0	0
7	65	29.5	2.5	27.0	729.00	0	0	0
TOTAL					5983.0	112	218	7

$$T_y = \sum_{i=1}^{g} (t_i^3 - t_i)$$

$$= (2^3-2)+(2^3-2)+(3^3-3)+(3^3-3)+(2^3-2)+(4^3-4)$$

$$+ (2^3-2)+(2^3-2)+(2^3-2)+(2^3-2)$$

$$= 7(2^3-2)+2(3^3-3)+(4^3-4) = 150.$$

Given these values for T_x (1932) and T_y (150), it is now possible to calculate r_s:

$$r_s = \frac{(N^3 - N) - 6\sum d_i^2 - (T_x + T_y)/2}{\sqrt{(N^3 - N)^2 - [(T_x + T_y)(N^3 - N)] + T_x T_y}}$$

$$= \frac{(30^3 - 30) - 6(5983) - (1932+150)/2}{\sqrt{(30^3 - 30)^2 - [(1932 + 150)(30^3 - 30)] + (1932)(150)}}$$

$$= \frac{-9969}{25913.69} = -.3847.$$

To determine the significance of r_s, if N is sufficient large (e.g., $N > 25$), a t statistic is obtained:

$$t = r_s \sqrt{\frac{N - 2}{1 - r_s^2}}$$

$$= -.3847 \sqrt{\frac{30 - 2}{1 - (-.3847)^2}} = -2.205.$$

This t statistic is approximately distributed as a Student's t with $df = N - 2$. The null hypothesis that $r_s = 0$ will be rejected if the obtained value of the t statistic is greater than the absolute critical value at the prestated one- or two-tailed α (e.g., $\alpha = .05$) and $df = (N - 2)$ (provided, of course, for one-tailed tests, that the t value is in the predicted direction).

Because the research hypothesis in our hypothetical study is directional in the negative direction and $\alpha = .05$, we will compare our calculated t value with a critical one-tailed t value with $df = 28$. These critical values can be obtained from any table of critical values for the t distribution available in most textbooks on statistics (e.g., Siegel & Castellan, 1988). This critical value, $|-1.70|$, is smaller than our calculated value, $|-2.205|$, and it *is* in the direction predicted. We can, therefore, reject the null hypothesis in favor of the alternative: There is a significant negative association between children's postintervention anxiety and their posthospital adjustment scores.

For small N, there are tables that provide the critical value of the r_s at set levels of alpha (e.g., Siegel & Castellan, 1988, Table Q, pp. 360-361). In this case, the null hypothesis of no association will be rejected if the obtained r_s is greater than the critical tabled value. When examining computer printouts, the null hypothesis of no relationship will be rejected if the obtained significance level (p) is less than α.

**Critical Assumptions of the Spearman
Rank-Order Correlation Coefficient**

1. *The two randomly selected variables, X and Y, are continuous variables with at least an ordinal level of measurement.*
2. *The two variables, X and Y, are paired observations.*

The data from our hypothetical study partially meet these assumptions. The two variables, children's anxiety and mothers' assessments of posthospital adjustment, are paired observations that are at least at the ordinal level of measurement. The data were not, however, drawn from a random sample.

Computer Commands

Figure 9.2 presents the SPSS for Windows dialogue box for the Spearman rho. This dialogue box was opened by highlighting *Statistics . Analyze* *Correlate . . . Bivariate* in the menus and selecting the *Spearman rho* statistic. Because we are undertaking a directional test, the *one-tailed* significance level was chosen.

Computer-Generated Output

Table 9.14 presents the SPSS for Windows syntax commands and computer printout for Spearman's rho. Note that although we used the same

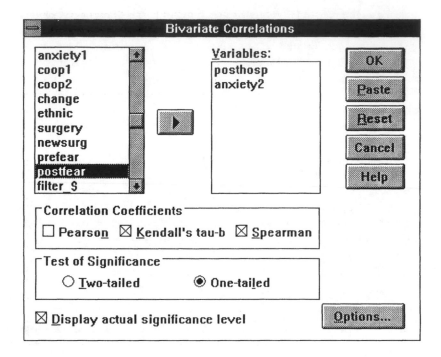

Figure 9.2. SPSS for Windows Computer Commands for the Spearman Rank-Order and Kendall's Tau-b Coefficients

dialogue box that would be used to generate the Pearson r, the syntax commands have indicated that we are using a nonparametric correlation coefficient (Table 9.14, ①).

The results of this analysis (②) indicated that, as we found earlier, there is indeed a negative relationship between children's postintervention anxiety and posthospital adjustment, $r_s = -.3847$. That is, as children's anxiety decreased, their posthospital adjustment scores increased. Because the one-tailed significance level, $p = .018$ (③), is less than our preset $\alpha = .05$, we can reject the null hypothesis and conclude that this inverse relationship is significant.

There is some discussion in the literature (Daniel, 1990; Strahan, 1982) as to whether r_s^2 can be used to assess the strength of relationship, because Spearman's rho addresses ranks and not actual values. Strahan (1982)

Table 9.14 SPSS for Windows Syntax Commands and Computer Printout for the Spearman Rank-Order Correlation and Kendall's Tau-b Coefficients

```
NONPAR CORR
  /VARIABLES=anxiety2 posthosp
  /PRINT= BOTH ONETAIL SIG  ①
  /MISSING=PAIRWISE .


- - -  S P E A R M A N   C O R R E L A T I O N   C O E F F I C I E N T S  -


POSTHOSP        -.3847  ②
            N(   30)
            Sig .018  ③

            ANXIETY2

(Coefficient / (Cases) / 1-tailed Significance)

- - - - K E N D A L L   C O R R E L A T I O N   C O E F F I C I E N T S  -
POSTHOSP        -.2831  ④
            N(   30)
            Sig .024  ⑤

            ANXIETY2

(Coefficient / (Cases) / 1-tailed Significance)
```

argues that r_s^2 is a good estimator of r^2 because, under the circumstances of a normal distribution, the magnitude of r_s is quite close to that of the Pearson r. He suggests, therefore, that r_s^2 is a reasonable nonparametric estimate of the percentage of variance in the dependent variable that can be explained by the independent variable.

In our hypothetical example, $r_s^2 = (-.3847)^2 = .1479$, which would suggest that approximately 14.79% of the variance in the children's post-hospital adjustment scores can be explained by their postintervention anxiety scores. According to the r^2 values presented in Table 7.6, this relationship is, at best, weak.

Presentation of Results

Table 9.15 is an example of a suitable presentation of the results of the Spearman rho correlation coefficient. Because the previously discussed

Table 9.15 Results of Tests of Association for the Point Biserial, Spearman
Rho, and Kendall's Tau Correlations

	Posthospital Adjustment	
	r	*p*
Gender		
Point biserial correlation	−.49	.007
Children's posttest anxiety		
Spearman rho	−.38	.018
Kendall's tau	−.28	.024

point biserial correlation and the upcoming Kendall's tau also used mothers' evaluations of children's posthospital adjustment as a dependent variable, these statistics are also presented. Note that the different tests could
be differentiated in footnotes to the table. The results of the Spearman rho
analysis could also be presented in the text as follows:

> The Spearman rank-order correlation coefficient was used to examine the
> extent to which children's postintervention anxiety scores were negatively
> associated with mothers' evaluations of children's posthospital adjustment.
> The results of this analysis ($r_s = -.38, p = .018$) indicated that children with
> greater postintervention anxiety were evaluated by their mothers to have
> poorer posthospital adjustment. The strength of this relationship ($r_s^2 = .148$)
> was weak in that only 14.8% of the variance in the children's posthospital
> adjustment scores could be explained by postintervention anxiety.

Advantages and Limitations of the Spearman
Rank-Order Correlation Coefficient

The Spearman rho correlation coefficient is an extremely useful, easily
calculated statistic that can be used when the assumptions of the parametric
Pearson correlation coefficient have not been met sufficiently. When the
assumptions underlying the Pearson *r* have been met, Spearman's rho is
91% as efficient as the Pearson *r* in rejecting the null hypothesis (Daniel,
1990; Siegel & Castellan, 1988). This means that if a correlation between
two continuous variables really exists in a population that has a bivariate

normal distribution, Spearman's rho will detect that relationship in 100 cases with the same significance as the Pearson r does in 91 cases (Siegel & Castellan, 1988).

Another advantage is that Spearman's rho closely approximates the numerical size of the Pearson r and its squared value is considered by some researchers to be a close nonparametric approximation of the coefficient of determination, r^2. A disadvantage of Spearman's rho is that it requires a larger sample size than does Kendall's tau to approximate a normal distribution and also does not have the same relationship to a partial correlation coefficient as does Kendall's tau.

Parametric and Nonparametric Alternatives to the Spearman Rank-Order Correlation

The Pearson product-moment correlation coefficient (Pearson r) is the parametric equivalent to the Spearman rho rank-order correlation coefficient. A nonparametric alternative to Spearman's rho is Kendall's tau coefficient. This statistic will be discussed in the next section.

Examples From Published Research

Cook, L., Smith, D. S., & Truman, G. (1994). Using Functional Independence Measure profiles as an index of outcome in the rehabilitation of brain-injured patients. *Archives of Physical Medicine and Rehabilitation, 75,* 390-393.

Coyle, L. A., & Sokop, A. G. (1990). Innovation adoption behavior among nurses. *Nursing Research, 39,* 176-180.

Wells, B. L., & Horm, J. W. (1992). Stage at diagnosis in breast cancer: Race and socioeconomic factors. *American Journal of Public Health, 82,* 1383-1385.

Kendall's Tau Coefficient

Kendall's tau coefficient was developed by Kendall (1938) as an alternative measure of association to Spearman's rho. It has been described as a measure of discrepancy or discordance between two continuous variables (Daniel, 1990; Siegel & Castellan, 1988). This coefficient is represented in research reports by various symbols (e.g., τ, T, or t) and also has been referred to as the Kendall rank-order correlation coefficient (Siegel & Castellan, 1988). In this text, the statistic will be referred to as *Kendall's tau* (τ).

An Appropriate Research Question for Kendall's Tau Coefficient

Kendall's tau coefficient may be used under the same conditions as Spearman's rho; that is, it can be used to examine the degree of association or dependence between two continuous variables. For example, in addition to Cramér's V, Metheny and colleagues (1989) used Kendall's tau to evaluate the strength of association between pH measurements and feeding tube placements. Marotz-Baden and Mattheis (1994) also used this coefficient, to examine the relationship between stress and integration into the extended family among daughters-in-law from two-generation farm families.

Because of its similarity to the Spearman rho coefficient, a research question similar to that posed in our hypothetical study for the Spearman rho will be used for Kendall's tau so that we can compare the results of these two nonparametric measures of association:

To what extent is there a negative relationship between children's postintervention anxiety scores and mothers' perceptions of their children's posthospital adjustment?

Null and Alternative Hypotheses

The null and alternative hypotheses for Kendall's tau are similar to those of Spearman's rho (Table 9.12). In our hypothetical study, for example, the null hypothesis for the above-stated research question would state that there is no relationship between children's posttest anxiety scores and mothers' perceptions of their children's posthospital adjustment. The alternative or research hypothesis would postulate an inverse or negative relationship between the two variables: Children with higher posttest anxiety will have poorer posthospital adjustment. Because the alternative hypothesis is directional, this is a one-tailed test.

Overview of the Procedure

Like Spearman's rho, Kendall's tau is based on the ranking of continuous data that are at least at the ordinal level of measurement. Under most conditions, the values of this coefficient range between -1 and $+1$, with the value of -1 suggesting a perfect inverse relationship between two continuous variables, 0 the lack of a relationship, and $+1$ a perfect direct relationship. Because of the differences in the way the two statistics are calculated, however, there are often discrepancies in their calculated values.

To calculate Kendall's tau, the values for variable X are first ranked in ascending order from lowest to highest (Daniel, 1990). These ascending values are considered to be in *natural order*. In Table 9.13, the Anxiety variable (X) is ranked from 3 to 7. Next, the values of variable Y within each value of X are also ranked from lowest to highest. For example, in Table 9.13, the values of the Adjustment variable (Y) have been ranked from 76 to 85 within the Anxiety score of 3 (③).

Each observation of Y is now compared to a Y value lying below it that does not share the same X value (e.g., the value of 76 when $X = 3$ is compared to the values of 75, 85, 89, 90, and so on that do not have an Anxiety value equal to 3). This pair of Y values is said to be in *natural order* (*concordant*) if the Y value below is larger than the first Y value (e.g., 76 and 85). The pair is in *reverse natural order* (*discordant*) if the Y value below is smaller than the Y value above (e.g., 76 and 75). The pair of Y values is considered to be *tied* if they share the same value (e.g., 75 and 75). The number of concordant, discordant, and tied Y pairs are obtained and summed across all values of Y to obtain the total number of Y pairs that are either in natural order (C) (④) or in reverse order (D) (⑤). To accommodate the presence of tied observations, the number of ties within the X and the Y observations are also recorded (⑥).

From these data, a Kendall tau coefficient (tau-a) is calculated. If there are no tied Y pairs, the following formula is used:

$$\tau_a = \frac{C - D}{n\,(n-1)/2}$$

where

C = the number of Y pairs in natural order

D = the number of Y pairs in reverse order

n = the total number of paired (X, Y) observations

$n(n-1)/2 = \binom{n}{2}$ = the total number of possible pairs of observations.

If the majority of Y pairs are in natural order (i.e., they are concordant), the value of Kendall's tau will be positive (i.e., $C - D > 0$). A positive value implies that as the ranking of the X variable increases (or decreases), the ranking of the Y variable follows suit. If the majority of the Y pairs are in reverse order (i.e., they are discordant), the value of Kendall's tau will be negative (i.e., $C - D < 0$); that is, increased ranks of the X variable are associated with decreased ranks of the Y variable. If the numbers of concordant and discordant pairs are equal, the value of Kendall's will be 0, implying no association between the two variables.

If there are no ties, the value of this Kendall's tau ranges between +1 (a perfect positive relationship) and −1 (a perfect negative relationship). If there are ties, the range of possible values for this coefficient is smaller because the number of concordant and discordant pairs will always be smaller than the total number of pairs (concordant + discordant + ties). To alleviate this problem, Kendall developed a second coefficient (tau-b) that takes into account the number of tied X and Y observations:

$$\tau_b = \frac{C - D}{\sqrt{[n(n - 1)/2] - T_x} \ \sqrt{[n(n - 1)/2] - T_y}}$$

where

$$T_x = \frac{1}{2} \sum t_x (t_x - 1)$$

$$T_y = \frac{1}{2} \sum t_y (t_y - 1)$$

t_x = the number of X observations that are tied at a given rank

t_y − the number of Y observations that are tied at a given rank.

With the numbers of ties now being taken into account, the value of Kendall's tau-b can range from −1 to +1, with higher absolute values indicating a stronger degree of association between the two variables. The null hypothesis of no association between X and Y will be rejected if the calculated value of Kendall's tau (a or b) exceeds its critical value at a prespecified level of alpha. Alternatively, when presented with a computer printout, the null hypothesis will be rejected if the generated p value for Kendall's tau is less than alpha (e.g., $\alpha = .05$).

There are several ways to determine the critical value of Kendall's tau. Several textbooks on nonparametric statistics (e.g., Daniel, 1990; Siegel & Castellan, 1988) present tables of critical tau values for small samples (the number of pairs of observations does not exceed $n = 40$). For larger samples, the following z statistic is approximately normally distributed with a mean of 0 and a variance of 1:

$$z = 3\tau \frac{\sqrt{n\,(n-1)}}{\sqrt{2\,(2n+5)}}$$

where

τ = the value of Kendall's tau

n = the number of pairs of observations.

The null hypothesis of no association between X and Y will be rejected if the absolute value of this calculated z statistic exceeds its absolute critical value (e.g., $|\pm1.64|$ for a one-tailed $\alpha = .05$) provided that, for a one-tailed test, it is in the direction predicted.

Calculating Kendall's Tau-b From Actual X and Y Values

Table 9.13 presented the actual values for the anxiety (X) and posthospital adjustment (Y) scores for 30 children in our hypothetical study. From these data, we obtained the squared differences in ranks d_i^2 (①) so that we could calculate Spearman's rho. We can also use these data to calculate Kendall's tau-b.

To calculate this statistic, the children's anxiety scores have first been ranked in natural order from lowest $(X = 3)$ to highest $(X = 7)$ (②), and the adjustment scores have also been ranked in natural order within each value of X. For example, for $X = 3$, the Y scores have been ranked 76, 76, 85, and 85 (③). The number of Y pairs *not included* in the particular X value that are in natural order, in reverse order, and tied are also presented. For $X = 3$ and $Y = 76$, for example, there were 16 concordant pairs (④), 9 discordant pairs (⑤), and 1 Y pair that shared the same value of 76 (⑥). The second value of $Y = 76$ when $X = 3$ (⑥) is not counted as a tie because the pair share the same X value (⑦).

Because there are so many tied X and Y values in this hypothetical study, it will be necessary to use Kendall's tau-b. We will need to calculate the number of tied observations for both X and Y. According to our formula for T_x:

$$T_x = \frac{1}{2} \sum t_x(t_x - 1)$$

$$= \frac{1}{2}[4(4-1)+7(7-1)+11(11-1)+6(6-1)+2(2-1)]$$

$$= \frac{1}{2}[4(3)+7(6)+11(10)+6(5)+2(1)] = \frac{1}{2}[196] = 98.$$

T_y is calculated by listing all 30 values of Y in ascending order and counting the number of ties (t_y) for each Y value:

$$T_y = \frac{1}{2} \sum t_Y(t_Y - 1)$$

$$= \frac{1}{2}[(2(1)+2(1)+3(2)+3(2)+2(1)+4(3)+2(1)+2(1)+2(1)+2(1)] = \frac{1}{2}(38) = 19.$$

Now it is possible to calculate Kendall's tau-b for our hypothetical study:

$$\tau_b = \frac{C - D}{\sqrt{[n(n-1)/2] - T_x} \ \sqrt{[n(n-1)/2] - T_y}}$$

$$= \frac{112 - 218}{\sqrt{[30(30-1)/2] - 98} \ \sqrt{[30(30-1)/2] - 19}}$$

$$= \frac{-106}{\sqrt{337} \ \sqrt{416}} = -.2831.$$

To determine whether the calculated tau value of $\tau_b = -.2831$ is sufficiently large to reject the null hypothesis, we could either consult a table of critical values for Kendall's tau (e.g., Daniel, 1990, Table A.22, p. S79) or calculate the z statistic outlined previously. Daniel's (1990) Table A.22 indicates that for $n = 30$ and a one-tailed $\alpha = .05$, the critical value for Kendall's tau is $-.218$. Because our value, $-.281$, is less than $-.218$, we can reject the null hypothesis in favor of the alternative.

The z statistic could be calculated as follows:

$$z = 3\tau \frac{\sqrt{n\,(n-1)}}{\sqrt{2\,(2n+5)}} = 3\,(-.2831)\,\frac{\sqrt{30\,(29)}}{\sqrt{2\,(60+5)}} = -2.197$$

The absolute value of the calculated z, $|-2.197|$, is greater than the one-tailed absolute critical value of z, $|-1.64|$, at $\alpha = .05$, and it is in the direction predicted. In fact, for $z = -2.197$, $p = .014$, which is considerably less than .05. The conclusion to be drawn is that there is a significant negative relationship between children's anxiety and their posthospital adjustment.

Critical Assumptions of Kendall's Tau Coefficient

As for Spearman's rho, there are not many critical assumptions associated with the Kendall tau coefficient. They are as follows:

1. *The randomly selected data are sets of paired observations (X, Y) that have been collected from the same subjects.*
2. *The two continuous variables, X and Y, are measured on at least an ordinal scale.*

Except for the issue of random selection, both of these assumptions have been met in our hypothetical study. The two variables, anxiety and posthospital adjustment, are paired observations collected from a single set of subjects. These two variables are continuous, having at least an ordinal level of measurement.

Computer Commands

The computer commands in SPSS for Windows for Kendall's tau coefficient are similar to those for Spearman's rho (Table 9.14). The dialogue box for this statistic is opened by clicking on *Statistics . . . Correlate . . . Bivariate* in the menu, selecting the continuous variables to be analyzed (*Anxiety2* and *Posthosp*), and clicking on *Kendall's Tau-b*. Because our alternative hypothesis was directional, we also requested a one-tailed significance test.

Computer-Generated Output

Table 9.14 presents not only the Spearman rho but also the computer-generated output for the Kendall tau-b coefficient obtained from SPSS for Windows. Note that, as indicated, the value for Kendall's tau-b (−.2831, ④) is smaller in absolute value than that for the Spearman rho (−.3834, ②) and has a larger p value: $p = 024$ for Kendall's tau-b (⑤) versus $p = .018$ for Spearman's rho (③). The conclusion regarding the null hypothesis is similar. Because this one-tailed p value, .028, is less than $\alpha = .05$, the null hypothesis will be rejected. The conclusion to be drawn is that, among the 30 children in our hypothetical study, there is an inverse relationship between postintervention anxiety and posthospital adjustment.

Presentation of Results

The presentation of the results for Kendall's tau-b is similar to that of the point biserial and Spearman rho correlations (Table 9.15). Similarly, a written summary of the analysis could be presented in the text as follows:

Kendall's tau-b coefficient was used to examine the extent to which children's postintervention anxiety scores were negatively associated with mothers' evaluations of children's posthospital adjustment. The results of this analysis ($\tau_b = -.28$, one-tailed $p = .024$) indicate that children with greater postintervention anxiety were evaluated by their mothers to have poorer posthospital adjustment.

Advantages and Limitations of Kendall's Tau Coefficient

Kendall's tau-b is used in circumstances similar to those for Spearman's rho to evaluate the degree of association between two continuous variables. Like the Spearman rho, it is useful when the assumptions underlying the parametric correlation coefficient, the Pearson product-moment correlation, have not been met. It has the advantage of having its distribution approach a normal distribution more quickly than does the Spearman rho distribution.

There are several statistics that have been generated from the basic Kendall tau-b formula that have extended its usefulness beyond a mere test of association. For example, Kendall's coefficient of concordance, W, is used not unlike the kappa coefficient to examine the degree of agreement among two or more raters concerning their rankings of objects or individuals (Daniel, 1990; Siegel & Castellan, 1988). Such a measure is especially useful for examining interrater reliability.

Unlike the Spearman rho, Kendall's tau-b can be generalized to a partial rank correlation coefficient that is very similar to the partial correlation coefficient obtained from the Pearson r. That is,

$$\tau_{xy.z} - \frac{\tau_{xy} - \tau_{xz}\tau_{yz}}{\sqrt{(1 - \tau_{xz}^2)(1 - \tau_{yz}^2)}}$$

This partial correlation coefficient examines the relationship between two variables, X and Y, while controlling for the effects of a third variable, Z. For example, in our hypothetical study, we might be interested in examining the relationship between the children's postintervention anxiety and posthospital adjustment, controlling for the effects of their preintervention anxiety. If we do not meet the assumptions of the parametric partial correlation coefficient using Pearson's r, we could use this nonparametric equivalent.

A disadvantage to Kendall's tau-b is the tediousness of calculating this statistic by hand, especially if the sample size is at all substantial. Fortunately, this is no longer a major consideration, given the availability of statistical computer packages.

Parametric and Nonparametric Alternatives to Kendall's Tau Coefficient

The parametric equivalent to Kendall's tau coefficient is the Pearson product-moment correlation (r). A nonparametric alternative to Kendall's tau is the Spearman rho rank-order correlation coefficient. Although the Spearman rho appears to be the more commonly used nonparametric measure of association for continuous variables in the health care literature, both of these statistics have the same asymptotic efficiency (.912) compared to the Pearson r (Stuart, 1954). That is, given a bivariate normal distribution, both the Spearman rho and Kendall's tau will reject the null hypothesis with 100 cases, at a significance level similar to that of the Pearson r with 91 cases. This suggests that, in terms of power (i.e., the ability to correctly reject the null hypothesis), both of these statistics are

Table 9.16 Nonparametric Correlation Coefficients That Would Be
Suitable With Variables of Specific Levels of Measurement

		Variable 2		
		Nominal	Ordinal	Interval/Ratio
Variable 1	Nominal	Phi (2×2) Contingency $(r \times c)$ Cramér V $(r \times c)$ Kappa		
	Ordinal	Rank biserial	Spearman's rho Kendall's Tau	
	Interval/ratio	Point biserial	Spearman's rho Kendall's Tau	Spearman's rho Kendall's Tau

nearly as powerful as the Pearson r given a normal distribution and can be more powerful than the Pearson r given a nonnormal bivariate distribution.

Examples From Published Research

Marotz-Baden, R., & Mattheis, C. (1994). Daughters-in-law and stress in two-generation farm families. *Family Relations, 43*, 132-137.
Metheny, N., Williams, P., Wiersema, L., Wehrle, M. A., Eisenberg, P., & McSweeney, M. (1989). Effectiveness of pH measurements in predicting feeding tube placement. *Nursing Research, 38*, 280-285.

Summary

In this chapter, we have examined six nonparametric measures of association between variables. Table 9.16 summarizes the statistics used and their expected levels of measurement.

The phi, Cramér's V, and kappa coefficients are used when both the independent and dependent variables are categorical. The point biserial correlation coefficient is useful when one of the two variables is dichotomous and the other is continuous. When the independent and dependent variables are both continuous (i.e., they are at least at the ordinal level of measurement), either Spearman's rho or Kendall's tau coefficient may be used. Three of these coefficients—the point biserial, phi, and Spearman's rho coefficients—are all special cases of a parametric statistic, the Pearson product-moment correlation coefficient (r).

There does not seem to be any definitive rule in the statistics literature as to which of the two rank-order coefficients, Spearman's rho or Kendall's tau, is preferred. It does appear from practice, however, that Spearman's rho is more commonly used in health care research (Pett & Sehy, 1996). This coefficient does have the advantage of being more close in size to the Pearson r and, therefore, r_s^2 has sometimes been used as an estimate of the amount of variance in the dependent variable that is explained by the independent variable. On the other hand, Kendall's tau can produce a partial rank-order correlation coefficient that is useful when the assumptions of its parametric counterpart have not been met.

Regardless of the method chosen, the researcher should be cautioned about proper interpretation of measures of association. As Gibbons (1985) has aptly warned, a significant test of association (parametric or nonparametric) provides no evidence of a causal relationship between two variables. Such a significant association could, in fact, be the result of another set of variables as yet unidentified. The existence of such a significant association, therefore, is a necessary but not sufficient condition for inferring causality, and significant results should be interpreted with caution.

10 Nonparametric Statistics: The Current State of the Art

The past nine chapters have examined a variety of nonparametric statistics that are useful under a number of different conditions. We have examined nonparametric statistics that are useful for assessing shapes of distributions (Chapter 4), tests of repeated measures for two or more samples (Chapters 5 and 6), tests of differences between two or more independent groups (Chapters 7 and 8), and measures of correlation between two variables (Chapter 9). The purpose of this final chapter is twofold: (1) to briefly summarize in table form the nonparametric statistics that have been examined in this text, the conditions under which they can be used, and their parametric counterparts; and (2) to identify those parametric statistical procedures for which there are currently no nonparametric alternatives in the popular statistical packages.

Currently Available Nonparametric Statistical Procedures

Table 10.1 provides a summary reference guide to the nonparametric statistical procedures reviewed in this text that are currently available in popular statistical computer packages. The reader may find this guide helpful when trying to make informed decisions concerning the most suitable nonparametric statistic to use given his or her data and purpose. The table also provides information as to which chapter the reader could turn to for further information concerning the nonparametric statistic of interest.

Although there may be additional nonparametric statistics that are offered by computer packages, these particular statistics were selected for

Table 10.1 Summary of the Nonparametric Statistical Tests Reviewed in This Text

	Single-Sample "Goodness of Fit"	Related Samples		Independent Samples		Tests of Association
		2 Measures	> 2 Measures	2 Levels	> 2 Levels	
Chapter Location	4	5	6	7	8	9
Dependent Variable: Nominal	binomial[1]; chi-square goodness-of-fit	McNemar	Cochran's Q	Fisher's[1]; chi-square	chi-square; Mantel-Haenszel	phi[1]; Cramér; Kappa
Ordinal/ Interval	Kolmogorov-Smirnov	Sign; Wilcoxon Signed Ranks	Friedman	median; Mann-Whitney U	median; Kruskal-Wallis	point biserial; Spearman's rho; Kendall's tau

1. Requires dichotomous variable.

their practical applicability to problems in health care research. They are also some of the most commonly used nonparametric statistics in the field.

For those readers who are trying to decide between a parametric and a nonparametric tests, Table 10.2 outlines the nonparametric alternatives to a particular parametric tests. Obviously, there are some nonparametric tests that do not have parametric alternatives because the variables being evaluated are at the nominal level of measurement (e.g., the chi-square tests and the phi and Cramér coefficients). In other instances (e.g., multiple regression), there are no nonparametric alternatives to the parametric tests that are currently available in the more popular statistical computer packages.

As has been emphasized throughout this text, it is extremely important that researchers be aware of the assumptions underlying the test (parametric or nonparametric) being considered and assess the extent to which their data meet those assumptions. No statistical test is powerful if its assumptions have been seriously violated. It is, therefore, the researcher's ethical

Table 10.2 Parametric Tests and Their Nonparametric Alternatives

Type of Problem	Parametric Test	Nonparametric Alternative
Repeated measures		
2 time periods	paired t test	Sign; Wilcoxon signed ranks
> 2 time periods	repeated-measures ANOVA ($1 \times c$)	Friedman
Independent samples		
2 groups	independent t test	median; Mann-Whitney U
> 2 groups	one-way ANOVA	median; Kruskal-Wallis
Measures of association	Pearson r	Spearman's rho; Kendall's tau

responsibility to ensure that tests of such assumptions have been carefully undertaken and reported.

The Limitations of Available Nonparametric Statistics

Although Table 10.2 suggests that there are a number of nonparametric alternatives to some parametric tests, there still remain major *lacunae* or holes in the nonparametric statistics field, especially concerning the computer availability of nonparametric statistics for more complex research designs. For example, although the Friedman test is a useful alternative when there are multiple observations for a single sample, there is, unfortunately, no nonparametric test in the statistical packages for use with repeated measures for two or more samples (e.g., repeated measures for an intervention and control group). There also is no nonparametric equivalent to the parametric ANCOVA, multiple regression, or other powerful multivariate parametric procedures available in most statistical computer packages. Many computer packages also do not include confidence interval estimates that are based on nonparametric estimates (a possible exception being MINITAB) (Gibbons, 1993).

A number of authors (e.g., Birkes & Dodge, 1993; Blair, 1981; Burnett & Barr, 1977; Harwell, 1991; Harwell & Serlin, 1988; Hettmansperger, 1984; Katz & McSweeney, 1980, 1983; Olejnik & Algina, 1984, 1985; Puri & Sen, 1969; Sawilowsky, 1990; von Eye, 1988; Zwick, 1985) have suggested nonparametric alternatives to the more widely used multivariate statistics. Unfortunately, these theoretical suggestions have not, for the most part, been translated into practical applications for the more popular statistical computer packages. Given that these statistics are rather complicated to compute, it is not suggested that the mathematically disinclined attempt these by hand.

Some recent authors have written about some potentially promising nonparametric multivariate techniques. Olejnik and Algina (1984, 1985) and Seaman, Algina, and Olejnik (1985), for example, have written extensively on the use of rank-transformed data in ANCOVA analyses. This approach, however, seems to have been met with mixed reviews concerning its robustness and power advantages over the parametric ANCOVA. A number of proposals have also been made with regard to the use of regression procedures based on the ranks rather than the actual values of data (e.g., Birkes & Dodge, 1993; Harwell & Serlin, 1989; Hettmansperger, 1984; Puri & Sen, 1985), but again it appears that only MINITAB provides needed nonparametric estimates and test statistics for the estimated regression line (Birkes & Dodge, 1993). It would appear, therefore, that the theoretical development of nonparametric statistics has superseded the practical development of computer applications that would facilitate the introduction of these nonparametric statistics into health care research.

When faced with data that do not meet the sometimes strict assumptions of multivariate statistical analyses, researchers can seek out simpler solutions. It may be possible to transform the dependent variable to achieve a more nearly normal distribution (see Chapter 3 for details). It might also be feasible to dichotomize the dependent variable and to consider using logistic regression in place of multiple regression. For repeated-measures ANOVA and ANCOVA, it might be advantageous to create change scores (e.g., Change = Time2 − Time1) and to undertake Kruskal-Wallis tests on these change scores. A disadvantage to change scores, however, is that they focus on *change*, not on scores that have been adjusted for the effects of a covariate. The results from these analyses could be subject to different outcomes and interpretations.

To conclude, when they are available, nonparametric statistics offer a feasible and potentially powerful alternative to the more restrictive parametric tests. As statistical computer packages catch up with theoretical developments in nonparametric statistics, the researcher can look forward to ever-expanding tools with which to reach informed decisions about research evidence.

Finally—The $64,000 Question

In the preface of this text, I indicated that I hoped that, at the conclusion of this text, the reader would be better able to answer the somewhat intoxicating question, "Does the Kolmogorov-Smirnov test *really* contain vodka?"

By now you must have ascertained that, indeed, this test does *not* contain alcohol. Some students of the K-S statistic, however, have reported that, when interpreting this statistic, it does help sometimes to have vodka near at hand—unless, of course, you prefer your vitamin C straight!

Appendix: QBASIC Routine for Subpartitioning an *r* × *k* Contingency Table

```
100 REM Test for the Chi-Square test for k independent samples
105 REM Routine for Subpartitioning an r x k Contingency Table
110 PRINT "PARTITIONING AN r x k CONTINGENCY TABLE INTO 2 x 2
    INDEPENDENT SUBTABLES"
120 PRINT "Originally presented in 1988 by N.J. Castellan, Jr."
125 PRINT "Updated for QBASIC by Dr. Marge Pett in 1996"
126 PRINT "Reprinted with permission from McGraw-Hill Publishers"
130 REM PRINT "For general r by k contingency tables"
140 PRINT : PRINT "You must enter the size of the contingency
    table"
160 INPUT "How many rows does your contingency table have"; R
170 INPUT "How many columns does it have"; K
180 DIM X(R, K), ROW(R), COL(K), E(R, K)
210 PRINT : PRINT "Please enter the data, cell by cell when
    requested to do so."
215 PRINT : PRINT "In addition, press the ENTER key after each
?"
220 FOR I = 1 TO R
230   FOR J = 1 TO K
240     INPUT X(I, J)
250     PRINT "Enter the data for the cell that occupies Row ";
        I; "Column"; J; : INPUT X(I, J)
260   NEXT J
270 NEXT I
280 REM Calculate marginal frequencies
290 FOR I = 1 TO R
300   FOR J = 1 TO K
```

```
310   ROW(I) = ROW(I) + X(I, J)
320   COL(J) = COL(J) + X(I, J)
330   N = N + X(I, J)
340  NEXT J
350 NEXT I
360 REM Find expected values and calculate chi-square (X2)
370 FOR I = 1 TO R
380   FOR J = 1 TO K
390   E(I, J) = ROW(I) * COL(J) / N
400   X2 = X2 + (X(I, J) ^ 2) / E(I, J)
410   NEXT J
420 NEXT I
430 X2 = X2 - N
440 PRINT : PRINT "Chi-square = "; X2; " with "; (R - 1) *
    (K - 1); "degrees of freedom."
450 REM Begin Partitioning Procedure
460 PRINT : PRINT "Partition cell(i,j) Chi-Square"
470 FOR J = 2 TO K
480   UR = X(1, J): UL = 0: LL = 0: LR = 0
490   FOR JJ = 1 TO J - 1: UL = UL + X(1, JJ): NEXT JJ
500   SR = 0: SC = SC + COL(J - 1)
510   FOR I = 2 TO R
520   UL = UL + LL
530   UR = UR + LR
540   LL = 0: FOR JJ = 1 TO J - 1: LL = LL + X(I, JJ): NEXT JJ
550   LR = X(I, J)
560   SR = SR + ROW(I - 1)
570   XT = N * (COL(J) * (ROW(I) * UL - SR * LL) - SC * (ROW(I)
      * UR - LR * SR)) ^ 2
580   XT = XT / (COL(J) * ROW(I) * SC * (SC + COL(J)) * SR *
      (SR + ROW(I)))
590   T = (R - 1) * (J - 2) + I - 1
600   PRINT USING "### ##:## ##.###"; T; I; J; XT
610   NEXT I
620 NEXT J
630 PRINT : PRINT "The critical chi-square value for df = 1,
    alpha = .05 is 3.84."
640 PRINT : PRINT "The critical chi-square value for df = 1,
    alpha = .01 is 6.64."
650 END
```

References

Most of the references below pertain to the use of statistics in general, but some are case studies of the use of particular statistics in practice. These references also are listed in the Examples From Published Research sections within the text.

Alkov, R. A., & Gaynor, J. A. (1991). Attitude changes in Navy/Marine flight instructors following an air crew coordination training course. *International Journal of Aviation Psychology, 1*, 245-253.

Allen, M., Weisman, C., Weedon, J., & Krasinski, K. (1993). Antepartum and intrapartum factors in perinatal HIV transmission. *International Conference on AIDS, 9*, 649.

Armitage, P., & Berry, G. (1994). *Statistical methods in medical research* (3rd ed.). London: Blackwell Scientific Publications.

Armstrong, G. D. (1981). Parametric statistics and ordinal data: A pervasive misperception. *Nursing Research, 30*, 60-62.

Bakeman, R., & Gottman, J. M. (1986). *Observing interaction: An introduction to sequential analysis.* Cambridge, UK: Cambridge University Press.

Balanda, K. P., & MacGillivray, H. L. (1988). Kurtosis: A critical review. *The American Statistician, 42*, 111-119.

Beach, E. K., Maloney, B. H., Plocica, A. R., Sherry, S. E., Weaver, M., Luthringer, L., & Utz, S. (1992). The spouse: A factor in recovery after acute myocardial infarction. *Heart & Lung: The Journal of Critical Care, 21*, 30-38.

Becker, P. T., Grunwald, P. C., Moorman, J., & Stuhr, S. (1993). Effects of developmental care on behavioral organization in very low birth weight infants. *Nursing Research, 42*, 214-220.

Bennett, B. M., & Underwood, R. E. (1970). On McNemar's test for the 2 × 2 table and its power function. *Biometrics, 6*, 339-343.

Bernard, A. (1992). The use of music as purposeful activity: A preliminary investigation. *Physical and Occupational Therapy in Geriatrics, 10*, 35-45.

Berry, K. J., & Mielke, P. W. (1987). Exact chi-square and Fisher's exact probability test for 3 × 2 cross classification tables. *Educational and Psychological Measurement, 47*, 631-636.

Bertoti, D. B. (1988). Effect of therapeutic horseback riding on posture in children with cerebral palsy. *Physical Therapy, 68*, 1505-1512.

Birkes, D., & Dodge, Y. (1993). *Alternative methods of regression.* New York: John Wiley.

Blackman, N. J., & Koval, J. J. (1993). Estimating rater agreement in 2 × 2 tables: Correction for chance and intraclass correlation. *Applied Psychological Measurement, 17*, 211-223.

Blair, R. C. (1981). A reaction to consequences of failure to meet assumptions underlying the fixed effects: Analysis of variance and covariance. *Review of Educational Research, 51,* 499-507.

Blair, R. C., & Higgins, J. J. (1985). Comparison of the power of the paired samples *t*-test to that of Wilcoxon's signed ranks test under various population shapes. *Psychological Bulletin, 97,* 119-128.

Blegen, M. A., Goode, C. J., Johnson, M., Maas, M. L., McCloskey, J. C., & Moorhead, S. A. (1992). Recognizing staff nurse job performance and achievements. *Research in Nursing and Health, 15,* 56-66.

Bond, C. A., & Monson, R. (1984). Sustained improvement in drug documentation, compliance, and disease control: A four-year analysis of an ambulatory care model. *Archives of Internal Medicine, 144,* 1159-1162.

Borenstein, M., & Cohen, J. (1988). *Statistical power analysis: A computer program.* Hillsdale, NJ: Lawrence Erlbaum.

Brennan, P. F., & Hays, B. J. (1992). The kappa statistic for establishing interrater reliability in the secondary analysis of qualitative clinical data. *Research in Nursing and Health, 15,* 153-158.

Brennan, R. L., & Prediger, D. J. (1981). Coefficient kappa: Some uses, misuses, and alternatives. *Educational and Psychological Measurement, 41,* 687-699.

Brown, G. W., & Hayden, G. F. (1985). Nonparametric methods: Clinical applications. *Journal of Clinical Pediatrics, 24,* 490-498.

Buck, J. L., & Finner, S. L. (1985). A still further note on Freeman's measure of association. *Psychometrika, 50,* 365-366.

Buckalew, L. W. (1983). Nonparametrics and psychology: A revitalized alliance. *Perceptual and Motor Skills, 57,* 447-450.

Burke, C. J. (1963). Measurement scales and statistical models. In M. H. Marx (Ed.), *Theories in contemporary psychology* (pp. 147-159). New York: Collier-Macmillan.

Burnett, T. D., & Barr, D. R. (1977). A nonparametric analogy of analysis of covariance. *Educational and Psychological Measurement, 37,* 341-348.

Burt-McAliley, D., Eberhardt, D., & van Rijswijk, L. (1994). Clinical study: Peristomal skin irritation in colostomy patients. *Ostomy Wound Management, 40,* 28-37.

Bustamante, E. A., & Levy, H. (1994). Sputum induction compared with bronchoalveolar lavage by Ballard catheter to diagnose Pheumocystis carinii pneumonia. *Chest, 105,* 816-822.

Cammu, H., & Van Nylen, M. (1995). Pelvic floor muscle exercises: Five years later. *Urology, 45,* 113-117.

Campbell, D. T., & Stanley, J. C. (1966). *Experimental and quasi-experimental designs for research.* Chicago: Rand McNally.

Campbell, J. P., Gratton, M. C., Salomone, J. A., & Watson, W. A. (1993). Ambulance arrival to patient contact: The hidden component of prehospital response time intervals. *Annals of Emergency Medicine, 22,* 1254-1257.

Chaplin, D., Deitz, J., & Jaffe, K. M. (1993). Motor performance in children after traumatic brain injury. *Archives of Physical Medicine and Rehabilitation, 74,* 161-164.

Cochran, W. G. (1952). The χ^2 test of goodness-of-fit. *Annals of Mathematical Statistics, 23,* 315-345.

Cochran, W. G. (1954). Some methods for strengthening the common χ^2 tests. *Biometrics, 10,* 417-451.

Cohen, J. (1960). A coefficient of agreement for nominal scales. *Educational and Psychological Measurement, 20,* 37-46.

Cohen, J. (1968). Weighted kappa: Nominal scale agreement with provision for scaled disagreement or partial credit. *Psychological Bulletin, 70,* 213-220.

Cohen, J. (1988). *Statistical power analysis for the behavioral sciences* (2nd ed.). Hillsdale, NJ: Lawrence Erlbaum.

Conger, A. J., & Ward, D. G. (1984). Agreement among 2×2 agreement indices. *Educational and Psychological Measurement, 44,* 301-314.

Conover, W. J. (1980). *Practical nonparametric statistics* (2nd ed.). New York: John Wiley.

Cook, L., Smith, D. S., & Truman, G. (1994). Using Functional Independence Measure profiles as an index of outcome in the rehabilitation of brain-injured patients. *Archives of Physical Medicine and Rehabilitation, 75,* 390-393.

Coyle, L. A., & Sokop, A. G. (1990). Innovation adoption behavior among nurses. *Nursing Research, 39,* 176-180.

Cramér, H. (1946). *Mathematical methods of statistics.* Princeton, NJ: Princeton University Press.

Crawford, J. D., & McIvor, G. P. (1985). Group psychotherapy: Benefits in multiple sclerosis. *Archives of Physical Medicine and Rehabilitation, 66,* 810-813.

Cronbach, L. J., Gleser, G. C., Nanda, H., & Rajaratmam, N. (1972). *The dependability of behavioral measurements: Theory of generalizability of scores and profiles.* New York: John Wiley.

Crozier, K. S., Graziani, V., Ditunno, J. F., & Herbison, G. J. (1991). Spinal cord injury: Prognosis for ambulation based on sensory examination in patients who are initially motor complete. *Archives of Physical Medicine and Rehabilitation, 72,* 119-121.

Damrosch, S. P., & Perry, L. A. (1989). Self-reported adjustment, chronic sorrow, and coping of parents of children with Down syndrome. *Nursing Research, 38,* 25-30.

Daniel, W. W. (1990). *Applied nonparametric statistics* (2nd ed.). Boston: PWS-Kent.

Dexter, F. (1994). Analysis of statistical tests to compare doses of analgesics among groups. *Anesthesiology, 81,* 610-615.

Dunn, O. J. (1964). Multiple comparisons using rank sums. *Technometrics, 6,* 241-252.

Eaves, R. C., & Milner, B. (1993). The criterion-related validity of the Childhood Autism Rating Scale and the Autism Behavior Checklist. *Journal of Abnormal Child Psychology, 21,* 481-491.

Ejaz, F. K., Folmar, S. J., Kaufmann, M., Rose, M. S., & Goldman, B. (1994). Restrain reduction: Can it be achieved? *The Gerontologist, 34,* 694-699.

Feingold, M. (1992). The equivalence of Cohen's kappa and Pearson's chi-square statistics in the 2×2 table. *Educational and Psychological Measurement, 52,* 57-61.

Feuer, E. J., & Kessler, L.G. (1989). Test statistic and sample size for a two-sample McNemar test. *Biometrics, 45,* 629-636.

Fleiss, J. L. (1971). Measuring nominal scale agreement among many raters. *Psychological Bulletin, 76,* 378-382.

Freeman, L. C. (1965). *Elementary applied statistics: For students in behavioral science.* New York: John Wiley.

Fromm, R. E., Levine, R. L., & Pepe, P. E. (1992). Circadian variation in the time of request for helicopter transport of cardiac patients. *Annals of Emergency Medicine, 21,* 1196-1199.

Gaddis, G. M., & Gaddis, M. L. (1994). Non-normality of distribution of Glasgow Coma Scores and Revised Trauma Scores. *Annals of Emergency Medicine, 23,* 75-80.

Gaither, N., & Glorfeld, L. (1983). An evaluation of the use of tests of significance in organizational behavior research. *Academy of Management Review, 10,* 787-793.

Gardner, P. L. (1975). Scaled statistics. *Review of Educational Research, 45,* 43-57.

Giannini, M. J., & Protas, E. J. (1991). Aerobic capacity in juvenile rheumatoid arthritis patients and healthy children. *Arthritis Care and Research, 4,* 131-135.

Gibbons, J. D. (1985). *Nonparametric methods for quantitative analysis.* Columbus, OH: American Sciences Press.

Gibbons, J. D. (1993). *Nonparametric statistics: An introduction.* Newbury Park, CA: Sage.

Gibbons, J. D., & Chakraborti, S. (1991). Comparisons of the Mann-Whitney, Student's *t,* and alternate *t* tests for means of normal distributions. *Journal of Experimental Education, 58,* 258-267.

Glazer, W. M., Morgenstern, H., & Doucette, J. (1994). Race and tardive dyskinesia among outpatients at a CMHC. *Hospital and Community Psychiatry, 45,* 38-42.

Goetz, A. M., Squier, C., Wagener, M. M., & Muder, R. R. (1994). Nosocomial infections in the human immunodeficiency virus-infected patient: A two year survey. *American Journal of Infection Control, 22,* 334-339.

Goodman, L. A. (1954). Kolmogorov-Smirnov tests for psychological research. *Psychological Bulletin, 51,* 160-168.

Graff-Radford, N. R., Godersky, J. C., & Jones, M. P. (1989). Variables predicting surgical outcome in symptomatic hydrocephalus in the elderly. *Neurology, 39,* 1601-1604.

Green, S. B. (1981). A comparison of three indexes of agreement between observers: Proportion of agreement, G-Index, and kappa. *Educational and Psychological Measurement, 41,* 1069-1072.

Haberman, S. J. (1984). The analysis of residuals in cross-classified tables. *Biometrics, 29,* 205-220.

Haddad, G. G., Jeng, H. J., Lai, T. L., & Mellins, R. B. (1987). Determination of sleep state in infants using respiratory variability. *Pediatric Research, 21,* 556-562.

Hair, J. F., Anderson, R. E., Tatham, R. L., & Black, W. C. (1995). *Multivariate data analysis with readings* (3rd ed.). New York: Macmillan.

Harsham, J., Keller, J. H., & Disbrow, D. (1994). Growth pattern of infants exposed to cocaine and other drugs in utero. *Journal of the American Diet Association, 94,* 999-1007.

Hartgers, C., van Ameijden, E. J., van den Hoek, J. A., & Coutinho, R. A. (1992). Needle sharing and participation in the Amsterdam Syringe Exchange program among HIV seronegative injecting drug users. *Public Health Reports, 107,* 675-681.

Harwell, M. R. (1988). Choosing between parametric and nonparametric tests. *Journal of Counseling and Development, 67*(1), 35-38.

Harwell, M. R. (1990). A general approach to hypothesis testing for nonparametric statistics. *Journal of Experimental Education, 58,* 143-156.

Harwell, M. R. (1991). Completely randomized factorial analysis of variance using ranks. *British Journal of Mathematical and Statistical Psychology, 44,* 383-401.

Harwell, M. R., & Serlin, R. C. (1988). An empirical study of a proposed test of nonparametric analysis of covariance. *Psychological Bulletin, 104,* 268-281.

Harwell, M. R., & Serlin, R. C. (1989). A nonparametric test statistic for the general linear model. *Journal of Educational Statistics, 14,* 351-371.

Hayes, L., Quine, S., & Bush, J. (1994). Attitude change amongst nursing students towards Australian Aborigines. *International Journal of Nursing Studies, 31,* 67-76.

Hays, W. L. (1994). *Statistics* (5th ed.). New York: Holt, Rinehart & Winston.

Hendrickse, W. A., Kusmiesz, H., Shelton, S., & Nelson, J. D. (1988). Five vs. ten days of therapy for acute otitis media. *Journal of Pediatric Infectious Diseases, 7,* 14-23.

Henry, S. B. (1991). Effect of level of patient acuity on clinical decision making of critical care nurses with varying levels of knowledge and experience. *Heart & Lung: The Journal of Critical Care, 20,* 478-485.

Herth, K. (1990). Relationship of hope, coping styles, concurrent losses, and setting to grief resolution in the elderly widow(er). *Research in Nursing & Health, 13,* 109-117.

Hettmansperger, T. P. (1984). *Statistical inference based on ranks.* New York: John Wiley.

Hildebrand, D. K. (1986). *Statistical thinking for behavioral scientists.* Boston: Duxbury.

Hinkle, D. E., Wiersma, W., & Jurs, S. G. (1994). *Applied statistics for the behavioral sciences* (3rd ed.). Boston: Houghton-Mifflin.

Hirji, K. F., Tan, S. J., & Elashoff, R. M. (1991). A quasi-exact test for comparing two binomial proportions. *Statistical Medicine, 10,* 1137-1153.

Hojat, M., Gonnella, J. S., & Xu, G. (1995). Gender comparisons of young physicians' perceptions of their medical education, professional life, and practice: A follow-up study of Jerrerson Medical College graduates. *Academic Medicine, 70,* 305-312.

Holland, B. S., & Copenhaver, M. D. (1988). Improved Bonferroni-type multiple testing procedures. *Psychological Bulletin, 104,* 145-149.

Holland, P. W., & Wainer, H. (1993). *Differential item functioning.* Hillsdale, NJ: Lawrence Erlbaum.

Hutchinson, S. W. (1990). Adolescent mothers' perceptions of newborn infants and the mothers' use of coping behaviors: A descriptive study. *Journal of the National Black Nurses' Association, 4,* 14-23.

Jacobson, N. S., & Moore, D. (1981). Spouses as observers of the events in their relationship. *Journal of Consulting and Clinical Psychology, 49,* 269-277.

Janicak, P. G., Davis, J. M., Gibbons, R. D., Erickson, S., Chang, S., & Gallagher, P. (1985). Efficacy of ECT: A meta-analysis. *American Journal of Psychiatry, 142,* 297-302.

Jeffery, P. K., Wardlaw, A. J., Nelson, F. C., Collins, J. V., & Kay, A. B. (1989). Bronchial biopsies in asthma: An ultrastructural, quantitative study and correlation with hyperreactivity. *American Review of Respiratory Disease, 140,* 1745-1753.

Jenkins, S. J., Fuqua, D. R., & Froehlc, T. C. (1984). A critical examination of the rise of nonparametric statistics. *Journal of Counseling Psychology, Perceptual & Motor Skills, 59,* 31-35.

Johnson, A. F. (1985). Beyond the technological fix: Outliers and probability statements. *Journal of Chronic Diseases, 38,* 957-961.

Kahn, H. A., & Sempos, C. T. (1989). *Statistical methods in epidemiology.* New York: Oxford University Press.

Kallenberg, W. C. M., Oosterhoff, J., & Schriever, B. F. (1985). The number of classes in chi-squared goodness-of-fit tests. *Journal of the American Statistical Association, 80,* 959-968.

Katerndahl, D. A. (1990). Comparison of panic symptom sequences and pathophysiologic models. *Journal of Behavior Therapy and Experimental Psychiatry, 21,* 101-111.

Katz, B. M., & McSweeney, M. (1980). A multivariate Kruskal-Wallis test with post hoc procedures. *Multivariate Behavioral Research, 15,* 281-297.

Katz, B. M., & McSweeney, M. (1983). Some nonparametric tests for analyzing ranked data in multi-group repeated measures designs. *British Journal of Mathematical and Statistical Psychology, 36,* 145-156.

Katz, N., & Sachs, D. (1991). Meaning ascribed to major professional concepts: A comparison of occupational therapy students and practitioners in the United States and Israel. *American Journal of Occupational Therapy, 45,* 137-145.

Kelly-Hayes, M., Wolf, P. A., Kase, C. S., Brand, F. N., McGuirk, J. M., & D'Agostino, R. B. (1995). Temporal patterns of stroke onset: The Framingham Study. *Stroke, 26,* 1343-1347.

Kendall, M. G. (1938). A new measure of rank correlation. *Biometrika, 30,* 81-93.

Knapp, T. R. (1990). Treating ordinal scales as interval scales: An attempt to resolve the controversy. *Nursing Research, 39,* 121-123.

Knapp-Spooner, C., & Yarcheski, A. (1992). Sleep patterns and stress in patients having coronary bypass. *Heart & Lung: The Journal of Critical Care, 21,* 342-349.

Koeslag, J. H., Schach, S. R., & Melzer, C.W. (1987). A reappraisal of the use of the phi coefficient in multiple choice examinations. *Medical Education, 21,* 46-52.

Kraemer, H. C., & Thiemann, S. (1987). *How many subjects?* Newbury Park, CA: Sage.

Krettek, J. E., Arkin, S. I., Chaisilwattana, P., & Monif, G. R. (1993). Chlamydia trachomatis in patients who used oral contraceptives and had intermenstrual spotting. *Obstetrics and Gynecology, 81,* 728-731.

Kruskal, W. (1988). Miracles and statistics: The casual assumption of independence. *Journal of the American Statistical Association, 83,* 929-940.

Kruskal, W. H., & Wallis, W. A. (1952). Use of ranks in one-criterion variance analysis. *Journal of the American Statistical Association, 47,* 583-621.

Lehman, R. S. (1991). *Statistics and research design in the behavioral sciences.* Belmont, CA: Wadsworth.

Levenson, J. L., Mishra, A., Hamer, R. M., & Hastillo, A. (1989). Denial and medical outcome in unstable angina. *Psychosomatic Medicine, 51*(1), 27-35.

Lezac, M. D., & Gray, D. K. (1984). Sampling problems and nonparametric solutions in clinical neuropsychological research. *Journal of Clinical Neuropsychology, 6,* 101-109.

Lilliefors, H. W. (1967). On the Kolmogorov-Smirnov test for normality with mean and variance unknown. *Journal of the American Statistical Association, 62,* 399-402.

Livingston, W. W., Healy, J. M., Jordan, H. S., Warner, C. K., & Zazzali, J. L (1994). Assessing the needs of women and clinicians for the management of menopause in an HMO. *Journal of General Internal Medicine, 9,* 385-389.

Ludbrook, J., & Dudley, H. (1994). Issues in biomedical statistics: Analyzing 2×2 tables of frequencies. *Australian and New Zealand Journal of Surgery, 64,* 780-787.

Lyman, B. (1982). The nutritional values and food group characteristics of foods preferred during various emotions. *Journal of Psychology, 112,* 121-127.

Mange, K. C., Marino, R. J., Gregory, P. C., Herbison, G. J., & Ditunno, J. F. (1992). Course of motor recovery in the zone of partial preservation in spinal cord injury. *Archives of Physical Medicine and Rehabilitation, 73,* 437-441.

Mann, H. B., & Whitney, D. R. (1947). On a test of whether one of two random variables is stochastically larger than the other. *Annals of Mathematical Statistics, 18,* 50-60.

Mann, J. M., Bila, K., Colebunders, R. L., Kalemba, K., Khonde, N., Bosenge, N., Nzilambi, N., Malonga, M., Jansegers, L., & Francis, H., et al. (1986). Natural history of human immunodeficiency virus infection in Zaire. *Lancet, 2,* 707-709.

Mantel, N. (1963). Chi-Square tests with one degree of freedom: Extensions of the Mantel-Haenszel procedure. *Journal of the American Statistical Association, 58,* 690-700.

Mantel, N., & Haenszel, W. (1959). Statistical aspects of the analysis of data from retrospective studies of disease. *Journal of the National Cancer Institute, 22,* 719-748.

Marotz-Baden, R., & Mattheis, C. (1994). Daughters-in-law and stress in two-generation farm families. *Family Relations, 43,* 132-137.

McAlindon, M. N., & Smith, G. R. (1994). Repurposing videodiscs for interactive video instruction: Teaching concepts of quality improvement. *Computers in Nursing, 12,* 46-56.

McCain, G. C. (1992). Facilitating inactive awake states in preterm infants: A study of three interventions. *Nursing Research, 40,* 359-363.

McMurdo, M. E., & Rennie, L. M. (1994). Improvements in quadriceps strength with regular seated exercise in the institutionalized elderly. *Archives of Physical Medicine and Rehabilitation, 75,* 600-603.

McNemar, Q. (1969). *Psychological statistics* (4th ed.). New York: John Wiley.

McSweeney, M., & Katz, B. M. (1978). Nonparametric statistics: Use and nonuse. *Perceptual and Motor Skills, 46,* 1023-1032.

Metheny, N., Williams, P., Wiersema, L., Wehrle, M. A., Eisenberg, P., & McSweeney, M. (1989). Effectiveness of pH measurements in predicting feeding tube placement. *Nursing Research, 38,* 280-285.

Munro, B. H., & Page, E. B. (1993). *Statistical methods for health care research* (2nd ed.). Philadelphia: J. B. Lippincott.

Myers, J. L., DiCecco, J. V., White, J. B., & Borden, V. M. (1982). Repeated measurements of dichotomous variables: Q and F tests. *Psychological Bulletin, 92,* 517-525.

Neave, H. R., & Worthington, P. L. (1988). *Distribution-free tests.* London: Unwin-Hyman.

Neter, J., Wasserman, W., & Whitmore, G. A. (1993). *Applied statistics* (4th ed.). Needham Heights, MA: Simon & Schuster.

Norusis, M. J. (1995a). *SPSS for Windows base system user's guide (Release 6.0).* Chicago: SPSS.

Norusis, M. J. (1995b). *SPSS 6.1 guide to data analysis.* Englewood Cliffs, NJ: Prentice Hall.

Novack, C. M., Waffarn, F., Sills, J. H., Pousti, T. J., Warden, M. J., & Cunningham, M. D. (1994). Focal intestinal perforation in the extremely-low-birth-weight infant. *Journal of Perinatology, 14,* 450-453.

Nunnally, J. C. (1978). *Psychometric theory* (2nd ed.). New York: McGraw-Hill.

Nunnally J. C., & Bernstein, I. (1994). *Psychometric theory* (3rd ed.). New York: McGraw-Hill.

Olejnik, S. F., & Algina, J. (1984). Parametric ANCOVA and the rank transform ANCOVA when the data are conditionally non-normal and heteroscedastic. *Journal of Educational Statistics, 9,* 129-149.

Olejnik, S. F., & Algina, J. (1985). A review of nonparametric alternatives to analysis of covariance. *Evaluation Review, 9,* 51-83.

Overall, J. E., & Hornick, C.W. (1982). An evaluation of power and sample size requirements for the continuity corrected Fisher exact test. *Perceptual and Motor Skills, 54,* 83-86.

Packer, T. L., Sauriol, A., & Brouwer, B. (1994). Fatigue secondary to chronic illness: Postpolio syndrome, chronic fatigue syndrome, and multiple sclerosis. *Archives of Physical Medicine and Rehabilitation, 75,* 1122-1126.

Patil, K. D. (1975). Cochran's Q test: Exact distribution. *Journal of the American Statistical Association, 70,* 186-189.

Pearson, K. (1920). Notes on the history of correlation. *Biometrika, 13,* 25-45.

Pedhazur, E. J., & Schmelkin, L. P. (1991). *Measurement, design and analysis: An integrated approach.* Hillsdale, NJ: Lawrence Erlbaum.

Pett, M. A., Lang, N., & Gander, A. (1992). Late life divorce: Its impact on family rituals. *Journal of Family Issues, 13,* 526-532.

Pett, M. A., & Sehy, Y. (1996). *The use and potential for misuse of parametric statistics in nursing research.* Unpublished manuscript, University of Utah, College of Nursing.

Philip, P. A., Ayyangar, R., Vanderbilt, J., & Gaebler-Spira, D. J. (1994). Rehabilitation outcome in children after treatment of primary brain tumor. *Archives of Physical Medicine and Rehabilitation, 75,* 36-39.

Pittinger, T. P., Maronian, N. C., Poulter, C. A., & Peacock, J. L. (1994). Importance of margin status in outcome of breast conserving surgery for carcinoma. *Surgery, 116,* 605-609.

Puri, M. L., & Sen, P. K. (1969). Analysis of covariance based on general rank scores. *Annals of Mathematical Statistics, 40,* 610-618.

Puri, M. L., & Sen, P. K. (1985). *Nonparametric methods in general linear models.* New York: John Wiley.

Ratner, P. A. (1995). Indicators of exposure to wife abuse. *Canadian Journal of Nursing Research, 27,* 31-46.

Richmond, T., Metcalf, J., & Daly, M. (1992). Cost of nursing care in acute spinal cord injury (SCI). *Heart & Lung: Journal of Critical Care Abstract, 21,* 293.

Robichaud-Ekstrand, S. (1991). Shower versus sink bath: Evaluation of heart rate, blood pressure, and subjective response of the patient with myocardial infarction. *Heart & Lung: The Journal of Critical Care, 20,* 375-382.

Rosen, M. G., Debanne, S. M., Thompson, K., & Dickinson, J. C. (1992). Abnormal labor and infant brain damage. *Obstetrics and Gynecology, 80*(6), 961-965.

Rothman, K. J. (1986). *Modern epidemiology.* Boston: Little, Brown.

Royeen, C. B., & Seaver, W. L. (1986). Promise in nonparametrics. *American Journal of Occupational Therapy, 40,* 191-193.

Ruttimann, U. E., & Pollack, M. M. (1991). Objective assessment of changing mortality risks in pediatric intensive care unit patients. *Critical Care Medicine, 19,* 474-483.

Savage, I. R. (1962). *Bibliography of nonparametric statistics.* Cambridge, MA: Harvard University Press.

Sawilowsky, S. (1990). Nonparametric tests of interaction in experimental design. *Review of Educational Research, 60,* 91-126.

Seaman, S. L., Algina, J., & Olejnik, S. F. (1985). Type I error probabilities and power of the rank and parametric ANCOVA procedures. *Journal of Educational Statistics, 10,* 345-367.

Sewerin, I. P. (1994). Clinical testing of the Ultra-Vision screen-film system for maxillofacial radiography. *Journal of Oral Surgery, Oral Medicine & Oral Pathology, 77,* 302-307.

Shavelson, R. J., & Webb, N. M. (1991). *Generalizability theory: A primer.* Newbury Park, CA: Sage.

Shure, D., & Astarita, R. W. (1983). Bronchogenic carcinoma presenting as an endobronchial mass. *Chest, 83,* 865-867.

Siegel, S., & Castellan, N.J. (1988). *Nonparametric statistics for the behavioral sciences.* New York: McGraw-Hill.

Singer, B. (1979). Distribution-free methods for non-parametric problems: A classified and selected bibliography. *British Journal of Mathematical and Statistical Psychology, 32,* 1-60.

Slakter, M. J. (1965). A comparison of the Pearson chi-square and Kolmogorov goodness-of-fit tests with respect to validity. *Journal of the American Statistical Association, 60,* 854-858.

Smiley, B. A., & Paradise, N. F. (1991). Does the duration of N_2O administration affect postoperative nausea and vomiting? *Nurse Anesthetist, 2,* 13-18.

Smirnov, N. V. (1939). Estimate of deviation between empirical distribution functions in two independent samples. *Bulletin Moscow University, 2,* 3-16. [Russian]

Snoey, E. R., Housset, B., Guyon, P., El Haddad, S., Valty, J., & Hericord, P. (1994). Analysis of emergency department interpretation of electrocardiograms. *Journal of Accident and Emergency Medicine, 11*, 149-153.

Spearman, C. (1904). The proof and measurement of association between two things. *American Journal of Psychology, 15*, 72-101.

Stevens, J. P. (1990). *Intermediate statistics: A modern approach*. Hillsdale, NJ: Lawrence Erlbaum.

Stevens, S. S. (1946). On the theory of scales of measurement. *Science, 103*, 677-680.

Stevens, S. S. (1951). Mathematics, measurement and psychophysics. In S. S. Stevens (Ed.), *Handbook of experimental psychology* (pp. 1-49). New York: John Wiley.

Stevens, S. S. (1968). Measurement, statistics, and the schemapiric view. *Science, 161*, 849-856.

Stolovitzky, J. P., & Todd, N. W. (1990). Head shape and abnormal appearance of tympanic membranes. *Otolaryngology, Head, & Neck Surgery, 102*, 322-325.

Strahan, R. F. (1982). Assessing magnitude of effect from rank-order correlation coefficients. *Educational and Psychological Measurement, 42*, 763-765.

Stroman, G. A., Stewart, W. C., Golnik, K. C., Cure, J. K., & Olinger, R. E. (1995). Magnetic resonance imaging in patients with low-tension glaucoma. *Archives of Ophthalmology, 113*, 168-172.

Stuart, A. (1954). The efficiencies of tests of randomness against normal regression. *Journal of the American Statistical Association, 51*, 285-287.

Tabachnick, B. G., & Fidell, L. S. (1996). *Using multivariate statistics* (3rd ed.). New York: HarperCollins.

Thornbury, J. M. (1992). Cognitive performance on Piagetian tasks by Alzheimer's disease patients. *Research in Nursing and Health, 15*, 11-18.

Uebersax, J. S. (1982). A generalized kappa coefficient. *Educational and Psychological Measurement, 42*, 181-183.

Van Lith, J. M., Pratt, J. J., Beekhuis, J. R., & Mantingh, A. (1992). Second trimester maternal serum immunoreactive inhibiting as a marker for fetal Down's syndrome. *Prenatal Diagnosis, 12*, 801-806.

Vlahov, D., Anthony, J. C., Celentano, D., Solomon, L., & Chowdhury, N. (1991). Trends of HIV-1 risk reduction among initiates into intravenous drug use 1982-1987. *American Journal of Drug and Alcohol Abuse, 17*, 39-48.

von Eye, A. (1988). Some multivariate developments in nonparametric statistics. In J. R. Nesselroade & R. B. Cattell (Eds.), *Handbook of multivariate experimental psychology* (2nd ed., pp. 367-398). New York: Plenum.

von Roenn, J. H., Bonomi, P. D., Gale, M., Anderson, K. M., Wolter, J. M., & Economou, S. G. (1988). Sequential hormone therapy for advanced breast cancer. *Seminars in Oncology, 15*, 38-43.

Walsh, J. E. (1946). On the power function of the sign test for slippage of means. *Annals of Mathematics and Statistics, 17*, 358-362.

Wampold, B. E., & Drew, C. J. (1990). *Theory and application of statistics*. New York: McGraw-Hill.

Watkins, A. J., & Kligman, E. W. (1993). Attendance patterns of older adults in a health promotion program. *Public Health Reports, 108*, 86-90.

Wechsberg-Wendee, M., Cavanaugh, E. R., Dunteman, G. H., & Smith, F. J. (1994). Changing needle practices in community outreach and methadone treatment. *Evaluation and Program Planning, 17*, 371-379.

Wells, B. L., & Horm, J. W. (1992). Stage at diagnosis in breast cancer: Race and socioeconomic factors. *American Journal of Public Health, 82,* 1383-1385.

Whiteman, K., Nachtmann, L., Kramer, D., Sereika, S., & Bierman, M. (1995). Effects of continuous lateral rotation therapy on pulmonary complications in liver transplant patients. *American Journal of Critical Care, 4,* 133-139.

Wilcox, R. R. (1992). Comparing the medians of dependent groups. *British Journal of Mathematical and Statistical Psychology, 45,* 151-162.

Wilcoxon, F. (1945). Individual comparisons by ranking methods. *Biometrics, 1,* 80-83.

Wolfer, J. A., & Visintainer, M. A. (1975). Pediatric surgical patients' and parents' stress responses and adjustment. *Nursing Research, 24,* 244-255.

Wright, L. J., & Carlquist, J. F. (1994). Measles immunity in employees of a multihospital health care provider. *Infection Control and Hospital Epidemiology, 15,* 8-11.

Zahniser, S. C., Gupta, S. C., Kendrick, J. S., Lee, N. C., & Spirtas, R. (1994). Tubal pregnancy and cigarette smoking: Is there an association? *Journal of Women's Health, 3,* 329-336.

Zimmerman, D. W. (1993). Increasing the power of rank tests by reducing the number of ranks. *Journal of Experimental Education, 61,* 271-277.

Zimmerman, D. W., & Zumbo, B. D. (1993). Relative power of the Wilcoxon test, the Friedman test, and repeated-measure ANOVA on ranks. *Journal of Experimental Education, 62,* 75-86.

Zwick, R. (1985). Nonparametric one-way multivariate analysis of variance: A computational approach based on the Pillai-Bartlett trace. *Psychological Bulletin, 97,* 148-152.

Index

Abuse, exposure rates for, 70
Adolescent mothers, 106
Agreement, interobserver, 237-248
Alpha level, 22-26
Alternative hypothesis, 14, 21-24. *See also*
     names of specific tests
Alternative tests, 277, 278(table). *See also*
     names of specific tests
Alzheimer's Disease, 88
Ambulatory care, 226
ANCOVA (analysis of covariance), 56, 278,
     279
Angina, 147
ANOVA (analysis of variance), 56, 181
    change scores and, 279
    Cochran's Q test and, 124
    Friedman test and, 132, 143-144
    median test and, 211
    See also Kruskal-Wallis one-way
        ANOVA
Artbuthnot, 15
Arthritis, 113
Association, tests of:
    biserial correlation, 249
    contingency coefficient, 166, 230
    Cramér's V, 166, 189-190, 232-237
    Kappa coefficient, 237-248
    Kendall's tau coefficient, 265-274
    phi coefficient, 164-166, 168, 190,
        225-232
    point biserial correlation, 248-255
    Spearman's rho, 255-265
    summaries of, 277(table), 278(table)
    tetrachoric correlation, 230
    See also Pearson product-moment
        correlation
Assumptions:
    for nonparametric tests, xi, 16-17,
        277-278

for parametric tests, ix, xi, 13-15, 17,
     277-278
    reporting on assessment of, 57
    robust characteristics and, 26
    See also Levels of measurement;
        Normality; Sample size; Variance;
        and names of specific tests
Asthma, 205
Asymmetric tables, 245-246
Asymmetry. *See* Skewness
Attitudes, changes in, 113
Autism rating scales, 226
Awake states, 132

Before and after tests. *See* Pretest posttest
     design
Beta error, 23. *See also* Type II error
Bimodal distributions, 44, 50. *See also*
     Skewness
Binomial distribution, 61-64
Binomial test:
    advantages of, 68
    alternatives to, 69
    applications of, 60, 69
    assumptions of, 64-65
    chi-square goodness-of-fit test and, 69
    directionality and, 60
    hypotheses for, 60
    levels of measurement for, 59-61, 64
    limitations of, 68-69
    McNemar test and, 100, 102
    presentation of results from, 68
    procedure for, 61-64
    research question for, 60
    sample size for, 62, 63
    sign test and, 106, 107
    SPSS for Windows for, 64-68
    usefulness of, 56, 68
    Wilcoxon signed ranks test and, 120

Biserial correlation, 249. *See also* Point
    biserial correlation
Block design, randomized, 122, 136
Bonferroni correction, 124, 140, 219
Boxplots, 49-50
Box's *M* test, 54
Brain injuries, 81, 113, 256
Brain tumors, 132
Breast cancer, 113, 158
Bronchogenic carcinoma, 123

Cancers, 113, 123, 158, 256
Cardiac patients, 81
Cardiologists, error rates of, 97
Categorical level. *See* Nominal level
Categories, collapsed. *See* Collapsed data
Cell influence, 163-164, 189, 190-197
Cell sizes:
    equality of, 14, 56
    minimum, 72-73, 157, 158, 168, 185, 202
    See also Sample size; Subgroups
Central Limit Theorem, 13-14
Central tendency. *See* Mean; Median; Mode
Cerebral palsy, 132
Change scores, 169-170, 279
Chi-square goodness-of-fit test:
    advantages of, 78
    alternatives to, 79
    applications of, 69-70, 79
    assumptions of, 71-73
    binomial test and, 69
    cell size for, 72-73
    directionality and, 71
    hypotheses for, 70-71
    K-S one-sample test and, 79, 86
    K-S two-sample test and, 93
    levels of measurement for, 69, 71, 72-73
    limitations of, 78
    normal distribution and, 78
    Poisson distribution and, 78
    popularity of, 18
    presentation of results from, 77
    procedure for, 71
    research question for, 69-70
    sample size for, 72-73
    SPSS for Windows for, 73-76
    usefulness of, 69, 78
Chi-square *k*-sample test:
    advantages of, 197
    alternatives to, 198
    applications of, 182-183, 198

    assumptions of, 185-186
    cell influence for, 189, 190-197
    cell size for, 185
    comparisons of, 223-224
    Cramér's *V* and, 189-190, 232
    directionality and, 184
    Fisher exact test and, 198
    hypotheses for, 183-184
    Kruskal-Wallis test and, 198, 223
    levels of measurement for, 182, 185, 197
    limitations of, 197-198
    Mantel-Haenszel test and, 198
    median test and, 198
    partition analysis in, 191-197
    phi coefficient and, 190
    popularity of, 18
    presentation of results from, 197
    procedure for, 184-185, 189-197
    research question for, 182-183
    residual analysis in, 190-191
    sample size for, 185
    SPSS for Windows for, 186-189
    strength of association and, 189-190
    usefulness of, 182, 197
Chi-square Mantel-Haenszel test, 198-204
Chi-square statistic:
    for Cochran's *Q* test, 125
    for Cramér's *V,* 233-234
    for Fisher exact test, 155
    for Friedman test, 135, 137
    for Kruskal-Wallis test, 214, 215-217,
        221
    for McNemar test, 99-100, 102
    for median test, 207
    for phi coefficient, 227
    Pearson, 161, 248
Chi-square two-sample test:
    advantages of, 167
    alternatives to, 168
    applications of, 157-158, 168-169
    assumptions of, 159-160
    cell influence for, 163-164
    cell size for, 157, 158, 168
    contingency coefficient and, 166
    Cramér's *V* and, 166
    Fisher exact test and, 156, 168
    hypotheses for, 158
    level of measurement for, 157, 160
    limitations of, 167
    phi coefficient and, 164-166, 168, 226
    popularity of, 18

presentation of results from, 167
procedure for, 158-159
residual analysis and, 163-164
sample size for, 157, 158, 163, 164, 166, 168
SPSS for Windows for, 160-164
strength of association and, 164-166
usefulness of, 157, 167
Chlamydia trachomatis, 226
Chronic fatigue syndrome, 212
Circadian rhythms, 81
Cochran's Q test:
    advantages of, 131
    alternatives to, 131
    applications of, 123-124, 131
    assumptions of, 127-128
    chi-square statistic for, 125
    directionality and, 124
    Friedman test and, 131
    hypotheses for, 124
    levels of measurement for, 96, 123, 127, 131
    limitations of, 131
    McNemar test and, 104
    presentation of results from, 130
    procedures for, 124-126
    research question for, 124
    sample size for, 125, 131
    SPSS for Windows for, 128-130
    usefulness of, 123, 131
Cognitive performance, 88
Collapsed data, 35, 61
    disadvantage of, 205
    for binomial test, 64-65
    for chi-square tests, 73, 76-77, 168, 183, 198, 204
    for K-S two-sample test, 89
    See also Partitioning
Colostomies, 113
Computer packages:
    limitations of, 278-280
    misconceptions about, 18
    reference guide to, 276-277
    See also SPSS for Windows
Computer program, for partitioning
    contingency tables, 193, 195, 281-282
Confidence intervals, 23
Confusion matrix, 242, 243
Contingency coefficient, 166, 230
Contingency tables:
    collapsing of cells in, 168

for chi-square k-sample test, 184
for chi-square two-sample test, 157, 168
for Cramér's V, 233
for Fisher exact test, 147, 157
for kappa coefficient, 240
for median test, 207
for phi coefficient, 227-228
in SPSS for Windows, 155
partitioning of, 191-197, 281-282
Continuity, Yates correction for, 163, 168
Contraception, 226
Convenience sampling, 22
Coping behaviors, 106, 249
Coronary artery bypass grafts, 249
Correlation, tests of, 29
    biserial correlation, 249
    Pearson. See Pearson product-moment correlation
    point biserial correlation, 248-255
    Spearman's rho, 255-265
    summaries of, 277(table), 278(table)
    tetrachoric, 230
    See also Association, tests of
Costs, of nursing care, 212
Counseling psychology, 18
Covariance, analysis of (ANCOVA), 56, 278, 279
Cramér's V:
    advantages of, 235-236
    alternatives to, 237
    applications of, 232-233, 237
    assumptions of, 234
    chi-square k-sample test and, 189-190, 232
    chi-square statistic for, 233-234
    chi-square two-sample test and, 166
    hypotheses for, 233
    levels of measurement for, 232, 234, 237
    limitations of, 234, 235-236
    Mantel-Haenszel test and, 203
    Pearson correlation and, 237
    phi coefficient and, 230, 237
    presentation of results for, 235
    procedure for, 233-234
    research question for, 232-233
    sample size for, 228, 230, 236
    Spearman's rho and, 237
    SPSS for Windows for, 234-235
    usefulness of, 232, 235-236
Criteria:
    for data characteristics, 58

for selecting parametric vs.
  nonparametric, 25-29
Critical care nurses, 97

Data:
  as missing values, 66-67, 74, 129, 137
  as outliers, 29, 39, 41, 49-52, 142
  as ties, 28, 171, 257-260, 267, 268
  checklist for assessment of, 58
  collapsed. *See* Collapsed data
  kurtosis and, 38-40, 53
  levels of. *See* Levels of measurement
  normality and, ix, 13, 35-54
  skewness of, 36-38, 39-40, 44, 53
  transformations of, 41, 52-54, 76, 83
  undefined, 83
  variance homogeneity and, 54
  See also Sample size
Data types. *See* Levels of measurement
Degrees of freedom, 71, 159, 184
DeMoivre, A., 15
Denial, impact of, 147
Dentistry, 123
Dependent data. *See* Related samples
Depression, 199
Detrended normal probability plots, 41, 44
Differences, between independent groups.
  See Independent samples
Differences, in sign. *See* Sign test; Wilcoxon
  signed ranks test
Directionality, 24. *See also names of
  specific tests*
Discordance, between variables, 265. *See
  also* Kendall's tau coefficient
Discriminant analysis, 198
Disease control, 226
Disease exposure rates, 70
Distribution-free tests, 15. *See also*
  Nonparametric tests
Distribution shape, 13, 16, 27-28, 29. *See
  also* Normal distribution; Normality
  binomial test for, 59-69
  chi-square test for, 69-79
  K-S Lilliefors test for, 44, 46-47, 83, 84
  K-S one-sample test for, 79-86
  K-S two-sample test for, 87-93, 175
  kurtosis and, 38-40, 53
  Mann-Whitney test for, 173-174, 175
  Shapiro-Wilks test for, 44, 46-47, 83
  skewness and, 36-38, 39-40, 44, 53
Divorce, 182

Down's syndrome, 88, 249
Drugs:
  documentation of, 226
  exposure to, 60
  injection of, 97, 170, 199
Dunn multiple comparison procedure,
  217-219

ECGs (electrocardiograms), 97
ECT (electroconvulsive therapy), 199
Education, 18, 113, 123, 205
Electrocardiograms (ECGs), 97
Electroconvulsive therapy (ECT), 199
Emotional states, 123
Equidistant intervals, 33, 34
Errors, in hypothesis testing, 22-26, 29, 55
Ethnicity, 60
Exclamation-point operator, 62
Exercise, 60, 113, 170
Exploratory studies, 23, 25
Extreme values, 49-50. *See also* Outliers
Eyeball test, 40-41

Factorial, 62
Failure to reject, 22-23, 29
Families, two-generation, 266
Feeding tubes, 232, 266
Fisher exact test:
  advantages of, 156
  alternatives to, 156
  applications of, 147-148, 156-157
  assumptions of, 151-152
  chi-square *k*-sample test and, 198
  chi-square statistic for, 155
  chi-square two-sample test and, 156, 168
  directionality and, 148
  hypotheses for, 148
  level of measurement for, 147, 151
  limitations of, 156
  McNemar test and, 104
  modified versions of, 156
  presentation of results from, 155
  procedure for, 148-151
  research question for, 148
  sample size for, 147, 149, 155, 156
  SPSS for Windows for, 147, 151, 152-155
  usefulness of, 147, 156
Fisher kurtosis coefficient, 39, 40
Fisher skewness coefficient, 36, 38, 39
Flight instructors, 113

Food preferences, 123
Framingham study, 70
Frequencies, focus on, 17
Friedman test:
    advantages of, 143
    alternatives to, 144
    ANOVA and, 132, 143-144
    applications of, 132, 144-145
    assumptions of, 136
    chi-square statistic for, 135, 137
    Cochran's Q test and, 131
    directionality and, 133
    hypotheses for, 133
    levels of measurement for, 132, 136
    limitations of, 143
    post hoc analyses for, 139-142
    power efficiency of, 143-144
    presentation of results from, 142-143
    procedure for, 133-135, 139-142
    research question for, 132
    sample size for, 135
    SPSS for Windows for, 135, 136-139
    usefulness of, 132, 143
    Wilcoxon tests for, 140-142
F test, 124, 135, 143-144, 181, 223. *See also*
    ANOVA

Generalizability theory, 248
Glasgow Coma Score, 70
Glaucoma, 113
Goodness-of-fit tests:
    binomial, 59-69
    chi-square, 69-79
    K-S Lilliefors, 44, 46-47, 83, 84
    K-S one-sample, 79-86
    K-S two-sample, 87-93
    popularity of, 18
    Shapiro-Wilks, 44, 46-47, 83
    summary of, 277(table)
    usefulness of, 56, 93-94
Grand median, 206-207
Grief resolution, 232
Group psychotherapy, 132
Groups, independent. *See* Independent
    samples
Growth patterns, of infants, 60

H_0. *See* Null hypothesis
H_a. *See* Alternative hypothesis
Health care research:

reports of, 17-18, 60
using binomial test, 60, 69
using chi-square goodness-of-fit test,
    69-70, 79
using chi-square k-sample test, 182-183,
    198
using chi-square two-sample test,
    157-158, 168-169
using Cochran's Q test, 123-124, 131
using Cramér's V, 232-233, 237
using Fisher exact test, 147-148, 156-157
using Friedman test, 132, 144-145
using kappa coefficient, 238-239, 248
using Kendall's tau, 266, 274
using Kruskal-Wallis test, 212-213, 223
using K-S one-sample test, 81, 86
using K-S two-sample test, 88, 93
using Mann-Whitney test, 169-171, 179
using Mantel-Haenszel chi-square test,
    199-200, 204
using McNemar test, 97, 105
using median test, 205-206, 211-212
using phi coefficient, 226, 231-232
using point biserial correlation, 249, 255
using sign test, 105-106, 112
using Spearman's rho, 256, 265
using Wilcoxon signed ranks test,
    113-114, 120-121
Health promotion, 212
Helicopter transport, 81
HIV (human immunodeficiency virus), 97,
    147, 158, 205
Holms stepdown procedure, 221
Homogeneity, of variance, 54, 252
Homoscedasticity, 14
Hormone therapy, 157
Human immunodeficiency virus (HIV), 97,
    147, 158, 205
Hypergeometric distribution, 149-150, 155
Hypotheses, 14, 20, 21-24, 29. *See also*
    names of specific tests
Hyrocephalus, 147

Immunoreactive inhibins, 88
Incontinence, 60
Independent samples:
    summaries of tests for, 277(table),
        278(table)
    two, 87-93, 146-180
    two, more than, 181-224
Independent samples, tests for:

chi-square for k samples, 182-198
chi-square for two samples, 157-169
Fisher exact, 147-157
Kruskal-Wallis one-way ANOVA, 212-223
K-S two-sample, 87-93
Mann-Whitney, 169-179
Mantel-Haenszel chi-square, 198-204
median, 204-212
summaries of, 277(table), 278(table)
Independent t test, 87, 93, 169, 178-179, 211
Infants, 60, 81, 106, 113, 132, 170
Inference. *See* Hypotheses
Innovation, 256
Interactive video instruction, 106, 123
Interobserver agreement, 237-248
Interval level, 13, 14, 27, 28, 33-34
    in summary of tests, 277(table)
    See also Levels of measurement
Intervals, equidistant, 33, 34
Intestinal perforation, 113
Inverse transformation, 53

Kappa coefficient:
    advantages of, 238, 247
    alternatives to, 248
    applications of, 238-239, 248
    assumptions of, 242
    asymmetric tables and, 245-246
    directionality and, 239
    estimated variance of, 241, 244
    hypotheses for, 239
    levels of measurement for, 242, 248
    limitations of, 247-248
    presentation of results from, 246-247
    procedure for, 240-242
    research question for, 238-239
    SPSS for Windows for, 241, 242-246
    standard error of, 241, 244
    usefulness of, 238, 247
Kendall, M. G., 265
Kendall's rank-order correlation coefficient.
    See Kendall's tau coefficient
Kendall's tau coefficient:
    advantages of, 272-273
    alternatives to, 273-274
    applications of, 266, 274
    assumptions of, 271
    directionality and, 266
    hypotheses for, 266
    levels of measurement for, 266, 271

limitation of, 273
Pearson correlation and, 273, 274
presentation of results from, 272
procedure for, 266-271
research question for, 266
Spearman's rho and, 265, 266, 269, 272, 273, 275
SPSS for Windows for, 271-272
usefulness of, 265, 272-273
Kendall's W test, 144
Kolmogorov, A. N., 87
Kolmogorov-Smirnov (K-S) Lilliefors test, 44, 46-47, 83, 84
Kolmogorov-Smirnov (K-S) one-sample test:
    advantages of, 86
    alternatives to, 86
    applications of, 81, 86
    assumptions of, 82-83
    chi-square goodness-of-fit test and, 79, 86
    directionality and, 81
    hypotheses for, 81
    levels of measurement for, 82-83, 86
    limitations of, 86
    Poisson distribution and, 79-80
    presentation of results from, 84
    procedure for, 81-82
    research question for, 81
    sample size for, 86
    SPSS for Windows for, 83-84
    usefulness of, 79-80, 86
Kolmogorov-Smirnov (K-S) two-sample test:
    advantages of, 92
    alternatives to, 93
    applications of, 88, 93
    assumptions of, 89
    chi-square goodness-of-fit test and, 93
    directionality and, 88
    hypotheses for, 88
    levels of measurement for, 87, 89
    limitations of, 92
    Mann-Whitney test and, 87, 93, 174, 175
    median test and, 87, 93
    presentation of results from, 92
    procedure for, 88-89
    sample size for, 89, 91-92
    SPSS for Windows for, 89-92
    usefulness of, 87, 92
Kruskal-Wallis one-way ANOVA:

advantages of, 222
alternatives to, 222-223
applications of, 212-213, 223
assumptions of, 214-215
change scores and, 279
chi-square *k*-sample test and, 198, 223
chi-square statistic for, 214, 215-217, 221
comparisons of, 223-224
Dunn procedure for, 217-219
Holms procedure for, 221
hypotheses for, 213
levels of measurement for, 212, 214
limitations of, 222
Mann-Whitney test and, 219-220
Mantel-Haenszel test and, 204, 223
median test and, 211, 223
post hoc comparisons for, 217-221
presentation of results from, 221-222
procedure for, 213-214, 217-221
research question for, 212-213
sample size for, 213-214
SPSS for Windows for, 215-217
usefulness of, 212, 222
K-S. *See* Kolmogorov-Smirnov
Kurtosis, 38-40, 53
K-W test. *See* Kruskal-Wallis one-way
ANOVA

Labor, abnormal, 81
Leptokurtic distributions, 38, 53
Levels of measurement, ix, 13, 14, 16,
27-28, 30-35
in summary of tests, 277(table)
interval, 13, 14, 27, 28, 33-34
nominal, 16, 27, 28, 30-32, 35
ordinal, 16, 27-28, 28, 32-33
ratio, 27, 28, 34
sample size and, 27-28, 56
See also names of specific tests
Levels of significance, 22-26
Levene test, 54
Lilliefors adjustment, 44, 46-47, 83, 84
Liver transplants, 170
Logistic regression, 70, 198
Log transformation, 53
Longitudinal studies, 96

Magnetic resonance imaging, 113
Mann, H. B., 169
Mann-Whitney test:

advantages of, 178
alternatives to, 178-179
applications of, 169-171, 179
assumptions of, 173-174
directionality and, 171
hypotheses for, 170-171
Kruskal-Wallis test and, 219-220
K-S two-sample test and, 87, 93, 174, 175
levels of measurement for, 169, 173
limitations of, 178
median test and, 179, 211
normal approximation for, 172
power efficiency of, 178
presentation of results from, 177-178
procedure for, 171-173
research question for, 169-171
sample size for, 172-173, 177, 178
SPSS for Windows for, 171, 174-177
usefulness of, 169, 178
Mann-Whitney *U* statistic, 169. *See also*
Mann-Whitney test
MANOVA (multivariate analysis of
variance), 56
Mantel-Haenszel chi-square test:
advantages of, 203
alternatives to, 204
applications of, 199-200, 204
assumptions of, 201-202
cell size for, 202
chi-square *k*-sample test and, 198
comparisons of, 223-224
Cramér's *V* and, 203
directionality and, 200
hypotheses for, 200
Kruskal-Wallis test and, 204, 223
levels of measurement for, 198, 201, 203
limitations of, 203-204
Pearson correlation and, 201
presentation of results from, 203
procedure for, 200-201
research question for, 199-200
sample size for, 201-202
Spearman's rho and, 204
SPSS for Windows for, 202-203
usefulness of, 198-199, 203
Mantel-Haenszel odds ratio, 198
Marital relationships, 239
Matched conditions:
summaries of tests for, 277(table),
278(table)
two, 95-121

two, more than, 122-145
McNemar test:
    advantages of, 104
    alternatives to, 104-105
    applications of, 97, 105
    assumptions of, 100-101
    binomial test and, 100, 102
    chi-square statistic for, 99-100, 102
    Cochran's Q test and, 104
    directionality and, 97-98
    Fisher exact test and, 104
    hypotheses for, 97-98
    levels of measurement for, 96, 97, 98,
        100-101, 104
    limitations of, 104
    presentation of results from, 103
    procedure for, 98-100
    research question for, 97
    sample size for, 99, 102
    sign test and, 105
    SPSS for Windows for, 101-103
    usefulness of, 96-97, 104
    Wilcoxon signed ranks test and, 105
Mean:
    in Wilcoxon signed ranks test, 119
    moments about, 36, 38
    outliers and, 49
    sample size and, 55
    skewness and, 36, 38
    test selection and, 58
Measles, 199
Measurement:
    definition of, 30
    levels of. See Levels of measurement
Measures, repeated:
    summaries of tests for, 277(table),
        278(table)
    two, 95-121
    two, more than, 122-145
Median:
    boxplots and, 50
    for Mann-Whitney test, 169
    grand, 206-207
    in Friedman test, 133, 143
    in Kruskal-Wallis test, 212
    in Wilcoxon signed ranks test, 116-117,
        119
    outliers and, 49
    skewness and, 36, 38
Median test:
    advantages of, 211

alternatives to, 211
applications of, 205-206, 211-212
assumptions of, 207-208
chi-square k-sample test and, 198
chi-square statistic for, 207
comparisons of, 223-224
directionality and, 206
hypotheses for, 206
Kruskal-Wallis test and, 211, 223
K-S two-sample test and, 87, 93
levels of measurement for, 204, 205, 207
limitations of, 211
Mann-Whitney test and, 179, 211
presentation of results from, 210-211
procedure for, 206-207
research question for, 206
SPSS for Windows for, 208-210
usefulness of, 204-205, 211
Medical education, 113, 205
Menopause, management of, 212
Mesokurtic distributions, 38
M-H test. See Mantel-Haenszel chi-square
    test
MINITAB, 279
Minorities, hospital use by, 60
Missing values, 66-67, 74, 129, 137
Mode, and skewness, 36
Moments, about the mean, 36, 38
Mortality rates, 70
MS (multiple sclerosis), 132, 212
M test, Box's, 54
Multiple choice questions, 226
Multiple observations. See Repeated
    measures
Multiple regression, 278
Multiple sclerosis (MS), 132, 212
Multivariate analysis of variance
    (MANOVA), 56
Multivariate outliers, 49, 51-52
Muscle exercises, 60
Music, and exercise, 113
Myocardial infarctions, 239

Natural order, of ranked data, 267
Needle practices, 97, 170
Neuropsychology, 18
Nominal level, 27, 28, 30-32
    collapsed data and, 35
    in summary of tests, 277(table)
    requirements for, 31
    See also Levels of measurement

Nonparametric tests:
    acceptance of, 17-18
    assumptions for, xi, 16-17, 277-278
    characteristics of, 16-17
    criteria for choosing, 25-29
    history of, 15
    hypotheses and, 20-29
    levels of measurement for, 16, 32, 33-34
    limitations of, 278-280
    misconceptions about, 18
    power of, 17, 26-27
    sample sizes for, 17, 24-25, 29
    summary of, 276-278
    types of, 19
    See also names of specific tests
Nonstatistical criteria, for test selection, 25,
    27-28
Normal distribution:
    binomial distribution and, 63-64, 107
    chi-square goodness-of-fit test and, 78
    history of, 15
    hypergeometric distribution and, 149-150
    Kappa coefficient and, 241
    Kendall's tau and, 269
    Mann-Whitney test and, 172
Normality, ix, 13
    assessment of, 35-47, 81
    data transformation and, 52-54
    kurtosis and, 38-40, 53
    sample size and, 55
    skewness and, 36-38, 39-40, 44, 53
    SPSS for Windows and, 35
    visual examination of, 40-44
Normal probability plots, 41, 44
Nosocomial infections, 147
Nuisance factor, 122
Null hypothesis, 14, 21-24, 29. *See also*
    names of specific tests
Nursing:
    assessments and, 239
    costs of, 212
    critical care, 97
    education for, 106, 113, 123
    in ambulatory care, 226
    innovation in, 256
    quality assurance in, 106
    recognition of, 132
    research reports on, 18, 60
Nursing homes, 182

Occupational therapy, 18, 106

Odds ratio, 198
One sample tests. *See* Single sample, tests
    for
One-tailed tests, 22, 24. *See also*
    Directionality
Oral contraception, 226
Ordinal level, 16, 27-28, 28, 32-33
    in summary of tests, 277(table)
    See also Levels of measurement; Rank
    ordered data
Organizational behavior, 17, 60
Osteoarthritis, 113
Otitis media, 147, 226
Outliers, 49-52, 142
    Fisher skewness coefficient and, 39
    nonparametric tests and, 29
    visual examination of, 41

Paired samples, tests for, 95-96
    McNemar, 96-105
    sign, 105-112
    Wilcoxon signed ranks, 112-121
    See also Correlation; Interobserver
    agreement; Repeated measures
Paired *t* test, 96, 112, 120
Panic symptoms, 81
Parametric tests:
    alternatives to, 277, 278(table)
    assumptions for, ix, 13-15, 17, 277-278
    characteristics of, 13-15
    criteria for choosing, 25-29
    hypotheses and, 20-29
    levels of measurement for, ix, 13, 14, 31,
        32, 33, 34
    power of, 17, 26-27
    sample sizes for, ix, 13, 24-25
    See also names of specific tests
Partitioning, of contingency tables, 191-197,
    281-282
Patient acuity, 97
Patient compliance, 226
PCP diagnosis, 97
Pearson, K., 15
Pearson chi-square statistic, 161, 248
Pearson product-moment correlation:
    Cramér's *V* and, 237
    Kendall's tau and, 273, 274
    Mantel-Haenszel test and, 201
    phi coefficient and, 165, 168, 230
    point biserial correlation and, 250, 252,
        255

Spearman's rho and, 256, 257, 262-263, 264-265
Pearson r. See Pearson product-moment correlation
Pearson skewness coefficient, 38, 39
Pelvic floor muscle exercises, 60
Peristomal skin irritation, 113
Pharmacists, 226
Phi coefficient:
  advantages of, 230
  alternatives to, 230
  applications of, 226, 231-232
  assumptions of, 228
  chi-square k-sample test and, 190
  chi-square statistic for, 227
  chi-square two-sample test and, 164-166, 168, 226
  Cramér's V and, 230, 237
  hypotheses for, 227
  level of measurement for, 225, 226, 228
  limitations of, 230
  presentation of results from, 230
  procedure for, 227-228
  research question for, 226
  sample size for, 230
  SPSS for Windows for, 228-230
  usefulness of, 225, 230
pH measurements, 232
Physical restraint reduction, 182
Physicians, error rates of, 97
Piagetian tasks, 88
Platykurtic distributions, 38, 53
Point biserial correlation:
  advantages of, 255
  alternatives for, 255
  applications of, 249, 255
  assumptions of, 251-252
  hypotheses for, 249-250
  levels of measurement for, 248-249, 251-252, 255
  limitations of, 255
  Pearson correlation and, 250, 252, 255
  presentation of results from, 254
  procedure for, 250-251
  research question for, 249
  SPSS for Windows for, 252-254
  usefulness of, 249, 255
Poisson distribution, 78, 79-80
Postpolio syndrome, 212
Power, of tests, 17, 26-27, 55
  efficiency and, 27, 112, 143-144

Pretest-posttest design, 20, 22
  tests for, 95-121
Probability plots, normal, 41, 44
Proportions:
  binomial test and, 59-68
  interobserver agreement and, 240-241
Psychology, 18
Psychotherapy, 132

Q statistic, 124-126. See also Cochran's Q test
Quality assurance, 106

Race, correlates of, 199
Randomized block design, 122, 136
Random sampling, xi, 13, 16
Rank ordered data, 16, 32, 52, 267. See also Ordinal level
Ratio level, 27, 28, 34. See also Levels of measurement
Recoded data. See Collapsed data; Transformations, of data
Regression, 51, 70, 198, 278, 279
Rejection, of null hypothesis, 22-24, 26-27, 29
Related samples:
  summaries of tests for, 277(table), 278(table)
  two measures of, 95-121
  two measures of, more than, 122-145
Related samples, tests for:
  Cochran's Q, 123-131
  Friedman, 131-145
  McNemar, 96-105
  sign, 105-112
  summaries of, 277(table), 278(table)
  Wilcoxon signed ranks test for, 112-121
Repeated measures:
  summaries of tests for, 277(table), 278(table)
  two, 95-121
  two, more than, 122-145
Research hypothesis, 14, 21-24. See also names of specific tests
Research reports, 17-18, 57. See also Health care research
Residual analysis, 161, 163-164, 190-191
Response times, 88
Reverse natural order, of ranked data, 267
Revised Trauma Score, 70

Rheumatoid arthritis, 113
Rho coefficient. *See* Spearman's rho
Robust tests, 17, 26
Rotation therapy, 170

Samples:
  convenience, 22
  independent, 87-93, 146-180, 181-224
  random, xi, 13, 16
  related, 95-121, 122-145
  single, 59-93
  summaries of tests for, 277(table),
    278(table)
Sample size:
  assessment of, 54-57
  levels of measurement and, 27-28, 56
  minimum, 13, 55-56
  normality and, 55
  parametric, ix, 13, 24-25
  parametric vs. nonparametric, 17, 24-25,
    29, 55-56
  power and, 27-28, 55, 112, 144
  subgroups and, 29, 56-57
  test selection and, 29, 56
  See also names of specific tests
Scales. *See* Levels of measurement
Sexual activity, 239
Shape. *See* Distribution shape
Shapiro-Wilks test, 44, 46-47, 83
Significance level, 22-26
Sign test:
  advantages of, 111
  alternatives to, 112
  applications of, 105-106, 112
  assumptions of, 108
  binomial test and, 106, 107
  directionality and, 106
  history of, 15, 105
  hypotheses for, 106
  levels of measurement for, 105, 106, 108,
    111
  limitations of, 111-112
  McNemar test and, 105
  power efficiency of, 112
  presentation of results from, 111
  procedure for, 106-108
  research question for, 106
  sample size for, 107
  SPSS for Windows for, 108-111
  usefulness of, 105, 111
  Wilcoxon signed ranks test and, 112, 120

Single sample, tests for:
  binomial, 59-69
  chi-square goodness-of-fit, 69-79
  K-S one-sample, 79-86
  summary of, 277(table)
Skewness, 36-38, 39-40, 44, 53
Skin irritation, 113
Sleep patterns, 81, 249
Smirnov, N. V., 87. *See also*
  Kolmogorov-Smirnov
Software. *See* Computer packages;
  Computer program
Sorted data. *See* Ordinal level
Spastic cerebral palsy, 132
Spearman rank-order correlation coefficient.
  See Spearman's rho
Spearman's rho:
  advantages of, 264-265
  alternatives to, 265
  applications of, 256, 265
  assumptions of, 261
  Cramér's *V* and, 237
  directionality and, 256
  hypotheses for, 256-257
  Kendall's tau and, 265, 266, 269, 272,
    273, 275
  levels of measurement for, 256, 261
  limitations of, 265
  Mantel-Haenszel test and, 204
  Pearson correlation and, 256, 257,
    262-263, 264-265
  presentation of results from, 263-264
  procedure for, 257-261
  research question for, 256
  sample size for, 265
  SPSS for Windows for, 261-263
  usefulness of, 255-256, 264-265
Spinal cord injuries, 212
SPSS for Windows, x, 35
  for binomial test, 64-68
  for box plots, 49-50
  for Box's *M* test, 54
  for chi-square goodness-of-fit test, 73-76
  for chi-square *k*-sample test, 186-189
  for chi-square two-sample test, 160-164
  for Cochran's *Q* test, 128-130
  for collapsing data, 64-65, 76
  for Cramér's *V,* 234-235
  for Fisher exact test, 147, 151, 152-155
  for Fisher kurtosis, 40
  for Fisher skewness, 39-40

for Friedman test, 135, 136-139
for kappa coefficient, 241, 242-246
for Kendall's tau coefficient, 271-272
for Kruskal-Wallis test, 215-217
for K-S Lilliefors test, 44, 46
for K-S one-sample test, 83-84
for K-S two-sample test, 89-92
for Levene test, 54
for Mann-Whitney test, 171, 174-177
for Mantel-Haenszel chi-square test,
   202-203
for McNemar test, 101-103
for median test, 208-210
for missing values, 66-67
for normality plots, 41, 44
for Pearson skewness, 39
for phi coefficient, 228-230
for point biserial correlation, 252-254
for regression analysis, 51
for Shapiro-Wilks test, 44, 46
for sign test, 108-111
for Spearman's rho, 261-263
for subgroups, 47
for transformations, 53
for Wilcoxon signed ranks test, 117-119
Sputum specimens, 97
Square root transformation, 53
Standard deviation, and test selection, 58
Standard error:
   for kurtosis, 40
   for skewness, 36, 39
Standardized residuals, 161, 164, 190-191
Standard normal distribution, 64, 108
Statistical criteria, for test selection, 25-27
Statistical inference. *See* Hypotheses
Statistical packages. *See* Computer packages
Stem-and-leaf plots, 47
Strength of association. *See* Association,
   tests of
Stress, 266
Stroke onset, 70
Student's *t*. *See* *t*-test
Subgroups:
   normality and, 47
   sample size and, 29, 56-57
   See also Partitioning
Substantive criteria, for test selection, 25,
   27-28

Tails. *See* Directionality
Tardive dyskinesia, 199

Tau coefficient. *See* Kendall's tau coefficient
Tests. *See names of specific tests*
Tetrachoric correlation, 230
Ties, in data, 28, 171, 257-260, 267, 268
Time periods:
   summaries of tests for, 277(table),
      278(table)
   two, 95-121
   two, more than, 122-145
Transformations, of data, 41, 52-54, 76, 83.
      See also Collapsed data
Traumatic brain injury, 113
t test:
   Dunn procedure and, 217-219
   independent, 87, 93, 169, 178-179, 211
   K-S two-sample test and, 93
   Mann-Whitney test and, 169, 178-179
   median test and, 211
   paired, 96, 112, 120
   point biserial correlation and, 251-252,
      255
   sign test and, 112
   Spearman's rho and, 260
   Wilcoxon signed ranks test and, 120
Two-sample tests. *See* Independent samples;
      Related samples
Two-tailed tests, 22. *See also* Directionality
Type I error, 22-23, 25-26, 29
Type II error, 22-23, 29, 55

Ultra-Vision screen, 123
Uniform distribution, 78, 79, 80
Univariate outliers, 49-51
U statistic, 172-173. *See also*
      Mann-Whitney test

Variability, between observers, 237
Variables, association between. *See*
      Association, tests of
Variables, level of. *See* Levels of
      measurement
Variance, 14, 29, 54. *See also* ANCOVA;
      ANOVA; MANOVA
Video instruction, 106, 123
Visual examination, of normality, 40-44

Whitney, D. R., 169. *See also*
      Mann-Whitney test
Wilcoxon, F., 113, 169

Wilcoxon-Mann-Whitney *U* test, 169-179.
    See also Mann-Whitney test
Wilcoxon signed ranks test:
    advantages of, 120
    alternatives to, 120
    applications of, 113-114, 120-121
    assumptions of, 115-117
    binomial test and, 120
    directionality and, 114
    Friedman test and, 140-142
    hypotheses for, 114
    levels of measurement for, 113, 115-116,
        131
    limitations of, 120
    McNemar test and, 105
    presentation of results from, 119-120
    procedure for, 114-115
    research question for, 114
    sample size for, 113

    sign test and, 112, 120
    SPSS for Windows for, 117-119
    usefulness of, 113, 120, 131
Wilcoxon *W* ranked sum statistic, 169. *See
    also* Mann-Whitney test

Yates continuity correction, 163, 168

Zero point, 34
z statistic:
    for binomial distribution, 63-64, 107
    for hypergeometric distribution, 149-150
    for kappa coefficient, 241-242, 244-245
    for Kendall's tau, 269, 271
    for Mann-Whitney test, 172-173
    for Wilcoxon signed ranks test, 115, 118

About the Author

Marjorie A. Pett, MStat, DSW, holds a joint appointment as Research Professor in the College of Nursing and in the Graduate School of Social Work at the University of Utah, Salt Lake City, having been a faculty member in the university since 1980. By her own admission, she is a "collector" of academic degrees: BA (Brown University), MS in sociology (University of Stockholm, Sweden), MSW (Smith College), and DSW (University of Utah). Her MStat in biostatistics (University of Utah) was obtained in 1991. Among other academic responsibilities, she has designed and taught graduate and undergraduate courses in research design and data management, parametric and nonparametric statistics. She has a strong interest in facilitating the practical application of statistics in the social, behavioral, and biological sciences, especially among practitioners in health care settings. She has tried to approach the teaching of statistics with humor and from a clinician's perspective. Prior to obtaining her doctorate in 1979, she worked for more than 15 years as a clinical social worker and family therapist in community mental health and private practice settings. Her research interests include divorce at all stages of the life cycle and observational studies of parent-child interaction in married and divorced families. She is the author of numerous research articles and chapters, and she has received several major funding grants from the National Institutes of Health.

3052578

Made in the USA